Conservation for Survival

KAI CURRY-LINDAHL

Conservation for Survival

An Ecological Strategy

LONDON · VICTOR GOLLANCZ LTD · 1972

ISBN 0 575 01549 7

Printed in Great Britain by
Lowe & Brydone (Printers) Ltd., London

FOREWORD

Man has brought an ecological crisis to the earth by destroying living environments. This crisis affects not only air, water, soil, plants, and animals, but ultimately man himself. Paradoxically the world's most intelligent species often behaves like the most foolish. No species except man and his domestic animals destroy the environment on which it depends.

Our planet is exposed to unbroken destruction as a consequence of continuous increase in the number of human beings. The larger populations grow, the greater the number who must starve, the more violent the struggle for space.

It is not likely that man will be able to colonize other planets within the foreseeable future. Planet earth is man's only hope for continuous existence. Therefore it should be of great concern to everyone to devote thought to the environmental problem of the earth, to try to make our planet biologically suitable for human life. To achieve that fitness, we would have to live what would be a truly normal life for our species. We cannot attain such a goal if we forget that we ourselves are biological organisms dependent on the natural environment.

Man's manipulation of environments almost always leads to an impoverishment of habitats and species. This in turn lowers the environmental quality even in areas where initially the biological simplification of the landscape is economically successful—for example, in crop monocultures. Man does not yet know how far he dare go in rendering the environment uniform, because there are no measurement standards available. In fact, in the ecological sense there are hardly two areas exactly alike. Despite this situation man simplifies the natural environment drastically and at a devastating rate.

Increased technological production coupled with increasingly larger human societies is using up resources rapidly without compensating for them. Instead man is in danger of drowning in his garbage of inorganic matter. In natural communities waste does not exist, because all living organisms become garbage, and this garbage forms the basis of the life and material resources of tomorrow. It is a perpetual system of life and death, death and life. Nothing in reality has been destroyed. Modern man has changed this simple but amazingly complex living system into a polluted system that threatens human welfare. The intensified man-made environmental pollution is creating cancers in the earth's living systems.

This book is concerned with the renewable natural resources—air, water, soil, vegetation and animals (including man), and the necessity of conservation of nature for survival. Obviously, the conservation of nature is of such fundamental importance for the world and its inhabitants that it should be the concern of governments to a far more wholehearted extent than has hitherto been the case. It is also obvious that were it not for the work of voluntary conservation organizations the world up to now would not have achieved much in the field of conservation.

Although the chapters of this book discuss natural resources such as air, water, soil, vegetation, animals, and man separately, it has been impossible to keep such subjects as, for example, erosion, pollution, or deforestation strictly to one single chapter. The renewable natural resources are so integrated that the effects of man's use of one of them immediately have repercussions in all the others. Therefore, the reader will find problems like, for example, pesticides discussed in all major chapters, but in this case the main treatment is found in the chapter dealing with animals.

Conservation of nature is applied ecology. It is a doctrine, or an ideology, based on biological facts. Its philosophy is the intelligent use of the earth's natural resources. Ecology could well become a new religion, based on indisputable facts.

This book, using examples and experiences from all continents and seas, tries to cover the basic conservation problems relative to the present world situation, in which man's conduct and treatment of the environment have led to an ecological crisis. This crisis is inescapable, and will lead inevitably to a catastrophe if man does not begin to think ecologically and then resolutely employ the power of his brain. He has to apply an ecological strategy.

How will today's tragic vandalism and shortsightedness be interpreted by tomorrow's generations? Natural values in the form of wild places, plants, and animals will be appreciated far more fully in the future than at the present. We already see indications of changing attitudes. If we irresponsibly allow the present mismanagement to continue, the people of the twentieth century will, despite their technological brilliance, go down in history as barbarians.

Not just buildings, art, electricity, cultivated fields, planted forests, and other man-made creations belong to our civilization. Free-living nature is also part of our heritage. If man destroys the last remnants of free nature, he forfeits the right to talk about himself in terms of civilization.

KAI CURRY-LINDAHL

ACKNOWLEDGMENTS

A special debt of gratitude is owed to Dr. Wendell G. Swank, Head of the Wildlife Division of the East African Agriculture and Forestry Research Organization, Muguga, Kenya, who read and commented upon the subchapters on Shifting Cultivation, Forest and Water Economy, Multiple Use of Grasslands and Forests, Productivity and Biomass, Deforestation, and Burning.

I am also grateful to Joseph R. Boldt, Jr., for the linguistic polishing of the manuscript, and to Lynn A. Ramsey for the careful copy editing.

CONTENTS

Conservation for Survival

CHAPTER ONE

Conservation for Survival

The human population of the world is increasing by 132 beings every minute, and this rate is accelerating. The chances for survival of other life forms continue to diminish as rapidly as their environments are destroyed by the rising tide of humanity. But like all other populations, man himself is dependent on the natural environment. We know from other species that uncontrolled growth in numbers sooner or later leads to a catastrophe—a population crash. In nature many barriers exist that prevent indefinite increase of populations. While man has bypassed many of the barriers that control his own numbers, he cannot escape them forever. In the long run his existence is based on the living world around him, and he is exposed to the same biological laws that govern animal populations.

The essential factor in seeking to conserve nature is to understand the function of the ecosystem, the living landscape, and the interrelationship between living organisms and their environment. The ecosystem is the basic functional unit of life, a marvelous pyramid of interactions among which capture and turnover of energy, production and productivity, biogeochemical cycling, especially of mineral nutrients, and so on are significant processes. A sound utilization of water, soils, plants, and animals must be suitably related to the long-term needs for maintaining and renewing these resources. The central question is: How, where, and when to harvest biological resources? To exploit them wisely does not necessarily mean a direct economic profit. The manifold riches of nature are much more than economic—they are also social, cultural, scientific, and aesthetic.

Since man is a trustee of natural resources for generations to follow, he has a moral obligation to administer this irreplaceable capital wisely and carefully. He must do it for the very survival of his own species. Endowed with the intellectual capacity to learn from previous catastrophic consequences of thoughtless exploitation, he has the gift—unique in nature—to foresee and plan the future.

Conservation of nature as a scientific discipline is virtually applied ecology. Though extremely complex in detail, ecology as a concept is simple: it deals with the interrelationships and interactions of all organisms—plants and animals, including man—with each other and with their living biological and physical environments. Unfortunately, for too long ecology has been the concern only of scientists. Most people are still ecologically illiterate. If modern man, particularly in town-dwelling populations, fails to realize even what ecology is, man as a species may fail to survive, may never realize his biological potential.

Ecology and biology are of great significance not only for the well-being and health of human individuals but also for the economy and prosperity of a society, a country, or a culture. Development schemes all over the world too often overlook this—a startling fact when one considers that a sound economic development is impossible without applying ecological principles. This is particularly true for many newly developing countries because they base their economy on agriculture—a land use that requires careful attention to all the ecological factors involved. The late Aldo Leopold, a great American ecologist and conservationist, admirably explained what ecology and conservation are about:

> A harmonious relation to land is more intricate, and of more consequence to civilization, than the historians of its progress seem to realize. Civilization is not, as they often assume, the enslavement of a stable and constant earth. It is a state of mutual and independent cooperation between human animals, other animals, plants, and soils, which may be disrupted at any moment by failure of any of them. Land-despoilation has evicted nations, and can, on occasion, do it again. As long as six virgin continents awaited the plough, this was perhaps no tragic matter–eviction from one piece of soil could be recouped by despoiling another. But there are now wars and rumours of wars which forestall the impending saturation of the earth's best soils and climates. It thus becomes a matter of some importance, at least to ourselves, that our dominion, once gained, be self-perpetuating rather than self-destructive.

The Need for a Global Campaign

Conservation of nature in a modern sense is the wise use of renewable natural resources. This means that man should try to reach a biological balance between his needs and nature's long-term capacity to satisfy them. "Renewable natural resources" include air, water, soils, plants, and animals. These are the resources essential to man's survival.

It is vital to understand the dependence of living organisms—plants and animals—on their habitats. A species is a product of a habitat, to which it constantly adapts through evolution, and is itself part of that habitat. Therefore the best way to preserve a species is to preserve the habitat. In many cases it is the only way.

An important part of modern nature conservation is just to ensure the existence of all types of habitats as well as of all plant and animal species. Man still knows too little about the living world around him. What today appears to be a most insignificant species in a river, on a savanna, or in a forest may tomorrow prove to play an extremely important environmental role or may provide products of great value for mankind. To preserve habitats and species is not merely to satisfy scientific curiosity or to save playgrounds for nature lovers. Man can benefit from the conservation of nature in many ways. In fact, to conserve the world's natural resources is to enhance the possibility of man's survival. If individuals, societies, and governments on the separate continents neglect this truth, they diminish that possibility.

Despite efforts in many countries—energetic actions by some governments and the magnificent work of some organizations, notably the International Union for Conservation of Nature and Natural Resources (IUCN)—the situation in nature conservation continues to deteriorate on a global scale and at a frightening rate. Destruction of air, waters, soils, plants, and animals in conjunction with human population growth has brought about such an increase of sterility over vast areas of water and land that man himself is threatened. This is not a bad dream; it is a likelihood that the evidence makes quite clear to every ecologist.

A general mobilizing of all forces against nature destruction is urgent and necessary. Wise use of natural resources is possible only when people understand and apply the rules and practices for conserving and managing them. Education is needed, but the difficulty is that the normal educational processes move too slowly to compete successfully with the accelerating human misuse of water and land. Nationwide campaigns for conservation must be started to arouse people to an awareness that man is as dependent on nature as an unborn child on its mother—that he is part of a biocommunity, and while he is its dominant member, in the long range perspective he is not its master—that nature contains forces man cannot override. Such a campaign presenting the philosophy and practice of conservation of nature should be coordinated with educational efforts at all levels. Its message must be channelled into national laws, international conventions, and global charters.

Air—Water—Soil—Vegetation—Animals—Man

In precivilization, man was just one member of the fauna, with no greater impact on the environment than other species. At that time there was no problem of conservation. This harmony still prevails where tribes do not cultivate the soil. The Pygmies in the Congo are a good example. They fish and hunt in the forest without destroying the habitat. When the animal populations decrease in the area, the nomadic Pygmies move to another part of the forest. The animal populations in the depleted area quickly regenerate. They remain prosperous, and so do the Pygmies, without breaking the harmony and the productivity of their habitat.

Nomadic tribes have generally not developed great civilizations. The latter have risen in areas where the natural resources have been virtually raped. Forests, fertile soils, and waters have disappeared. This has happened on every continent permanently inhabited by man. Despite his intelligence and marvelous ability to create and invent, man has continuously plundered his environment instead of living in peace with it. This is in contrast to all wild animal species and constitutes a terrible defect in human society. Man is destroying his own habitat, his basis of subsistence —he is cutting his own throat.

No civilization can survive in a ruined environment. History has taught this lesson many times. It will be repeated in the future unless the capital robbed from nature is repaid so that it again yields its former dividends. Hence we must aim at a colossal restoration of nature all over the world where the waters, soils, vegetation, and animal populations have been destroyed. This is an enormous but absolutely essential task because no other civilization prior to ours has carried the destruction of natural resources so far.

Every living organism requires water in order to live. To man water is as important as the air he breathes. Nevertheless, he has destroyed water resources all over the world in many ways and on a gigantic scale. The destruction is manifest in the ground water table, lakes, marshes, the running waters, estuaries, and even the seas.

Almost everywhere the loss of water resources has been due to the destruction of soils that have gradually become vulnerable because of overcultivation, agricultural malpractice, overgrazing, overtrampling, disappearance of forests, and loss of vegetation cover. Erosion by wind and water has removed fertile soils, of which the main part has been washed out into the sea. This still continues.

The vegetation plays an important role in stabilizing the water cycle and preserving the soil and in turn cannot exist without them. It facilitates evaporation and keeps the soil porous. Humus and soil filter melting snow and rainwater, resulting in flows of crystal-clear streams and springs. Important salt nutrients both from the air and the rains, essential for vegetal and animal life, also flow to the soil and the subsoil water. Forests accumulate, store, and distribute water to surrounding areas during dry periods. These are just instances of a very complex pyramid of interrelationships. When man destroys one or several of the components, the pyramid collapses.

If the forest vanishes and the earth's vegetation goes, the fertile soil layer disappears. When the vegetation dies, so do the animals. When the country thus grows sterile, the civilization existing there cannot survive. Many of today's deserts lie as silent memorials to once glorious civilizations that two thousand to three thousand years ago were based on extremely fertile lands supporting important animal and human populations.

The wild fauna are also a vital natural resource. Irreplaceable biological treasures—chiefly mammals and birds—are gone forever because their habitats were altered or completely destroyed, or the animals were wiped out by direct killing, without thought of their value as a natural resource.

Looking back on what man has achieved negatively on this globe of ours during a few centuries, it becomes evident that human beings must foreswear self-regard as lords of creation. Such a philosophy is a self-deception. Man should try to be master of himself by utilizing his intellectual ability to repair the damage he has inflicted on nature, to manage the present resources soundly, and to plan their utilization on a sustained yield basis.

The Living Landscape

It has been said that the human species is a disease of the soil. I would like to expand that notion to include in the areas of infection other elements that form a living, flourishing landscape. Until clumsy man entered the scene, everything functioned in a rational way with optimal productivity in every kind of habitat, aquatic as well as terrestrial. Man not only ruined the soil, he burned and slashed the vegetation. Where he had not already drained the water, he polluted it. He exterminated animal life, forgetting that wild animals can flourish where domestic stock will perish. He substituted dead lands for living landscapes. Now the reaction to this misuse has come, but only in cries of warning from the few, without any global effect. Except for local progress in the form of careful farming and reforestations, severe destruction of water and land goes on. Air and water pollution, erosion, eradication of forests and other vegetation, slaughter of animals, and overexploitation of marine resources reflect the contemporary human attitude toward natural environments.

Man's technical skills have eliminated the broad spectrum of habitat diversity that once was characteristic of the living landscapes. This diversity was one of the many factors that guaranteed biological and ecological health to all the landscape inhabitants. Today uniformity and monocultures have replaced the old diversity. Landscape similarity is a feature of huge areas not only within one country but from one continent to another. Such uniformity with great expanses of monocultures may at first be very profitable for the owners. But because lack of diverse ecological niches, with their wide variety of plant and animal species, favors the proliferation of diseases and parasites, in the long run and with few exceptions the environment will deteriorate. Unchecked by their natural controller, the pests spread rapidly, forcing man to undertake a chemical warfare against them and plant sicknesses. The sprays, dusts, and other ammunition of this warfare so poison the environment that the resulting biological impoverishment and accumulation of persistent toxic chemicals endanger not only plants and animals but man himself.

But there is still hope. The renaissance of a diverse living landscape, with its whole natural range of habitats, plant and animal, can be achieved when men come fully to understand that such environments will enable them to get profits from their crops.

Man's Place in the Biocommunity

Within a span of a few centuries man's role in changing the natural environments has been nothing less than cataclysmic. Geological forces and ice ages have achieved somewhat comparable havoc, but they permitted nature to reacquire what had been altered or destroyed. Nature will likewise reassert control if and when man obliterates himself. But by the same token man can avoid obliteration by working with nature instead of against it. He cannot do much concerning such forces as the sun's energy, winds, and tides, but he can contribute immensely to keeping the air and the water clean, the soil fertile, the vegetation diverse, and the animals abundant. He must regard himself as a biological being, an important part of the metabolism, energy flow, and function of the environment, and utilize his intelligence to maintain and wisely manage the renewable natural resources.

No single formula can tell him how to do all this—especially since the living landscape is a changing, dynamic community. Ecological research will give the answer to many questions and indicate what measures should be taken.

Humanity needs a global natural resources policy. The conservation of nature and natural resources, including habitats and wildlife, is a matter of such fundamental importance for the well-being of man that the matter should be of concern to governments of all nations. It is of the utmost importance for every country that conservation of nature be represented in the top administrative echelons of the government. Conservation should have as its national spokesman an individual at cabinet level whose entire task is to conserve and manage natural resources and to coordinate their utilization.

But it is necessary to look beyond national boundaries in order to avoid ecological crises on a giant scale. A powerful intergovernmental board exercising global supervision of a program for conserving and managing natural resources is badly needed—however politically unrealistic it may presently seem. The day will come when governments will think ecologically and realize that a world-wide conservation policy is necessary for human survival and the continued existence of nations.

CHAPTER TWO

The Air

The growth of cities and industries is generally considered indicative of progress and prosperity. But while the industrial society enables man to raise his economic standards, it also brings a deterioration of environmental health. The air that man must breathe is in many areas heavily polluted. The quality of the air—our most utilized natural resource—determines each individual's physical well-being or misery. We do not yet know the extent to which the inhaling of invisible poisonous waste weakens human health and shortens the life span, but the evidence is accumulating.

Air pollution is produced from many sources, and except for volcanic eruptions man is responsible for them Exhaust fumes from industries, smoke from domestic chimneys and urban dumps, dangerous gases from hundreds of thousands of gasoline-driven cars within just one small area, and so forth—these are some of the pollutant waste products constantly discharged in incredible quantities into the air. It is estimated that in Paris 47 percent of the atmospheric pollution is produced by gases from motor vehicles, 33 percent from domestic chimneys, and 20 percent from industries. However, in many cities the greatest source of air pollution—accounting for about 75 percent—is domestic fires. In a city like Los Angeles, with more than three million cars daily emitting an enormous volume of exhaust fumes, the motor vehicle contributes heavily, together with the local climate, to the most severe urban air pollution in the world.

The continuous pouring of poisonous fumes and gases into the air becomes particularly dangerous to all life forms where local atmospheric conditions create stagnant air masses, fogs, and temperature inversions. These accumulate the released airborne waste and build up a smog. Such atmospheric events are first links in what we may call the air pollution chain, which ends in our lungs or in those of the animals, or is absorbed by plants, poisoning and sometimes killing whatever life it is taken up by. It also cuts off the sun, reducing the effect of its rays on the photosynthesis process. Even the climate may be changed.

If human city dwellers could see or realize the amount of poisonous, dirty waste products they inhale daily, they would probably find their urban abode intolerable. At the least, they would be seriously shocked. Many would ask: Must we really pay this price for urban comfort—and for civilization?

Most people in larger U.S. cities regularly inhale air that is dirty with 10,000 particles per cubic cm. Scientists categorize a cubic cm. of air with 2,000 or less particles as clean air. A lung specialist in a New York City hospital states that there is little doubt that living in a polluted area is like taking a few years off one's life.

Atmosphere

The lower part of the air is normally thoroughly mixed and of rather uniform composition except in and above cities, industrial centers, big fires, volcanoes, etc. However, winds carry polluted air great distances, and concentrations and deposits of poisonous substances may occur far from the sources. In Europe, for example, sulphur spewed out from industrial areas in Germany and Great Britain damages neighboring countries and reaches as far as Scandinavia. Thus prevailing wind patterns may carry pollution to areas where people have no control over the poisons afflicting them. Radioactive fallout is also wind-transported and can contaminate land, water, plants, and animals thousands of miles away from the source. This problem is indeed international. Recent layers of snow and ice in Antarctica contain measurable amounts of lead. . . .

The major gaseous constituents of normal air are 78 percent nitrogen and 21 percent oxygen. The carbon dioxide content usually is not more than 0.03 percent. All living organisms must have direct or indirect access to oxygen at all times. The cycles of oxygen, nitrogen, and carbon are complex processes of basic importance for a living environment. Frequent disturbances caused by air pollution may create a serious imbalance in habitats and may upset entire ecosystems.

An even more serious threat to the air phase of living environments, including man's, is an upset of the photosynthesis processes brought on by destruction of the vegetation, both in water and on land, that replenishes the atmospheric oxygen supply. The green algae in the oceans, tiny marine plants that produce not less than 70 percent * of the atmospheric oxygen, are continuously affected by pollution from toxic chemicals that continually contaminate the ocean waters. For example, only one milligram of DDT in one ton of water—one part in ten million—is enough to reduce the ability of some green algae (planktonic diatoms) to produce oxygen. Today DDT has been detected in all waters of the globe—even off the coasts of Antarctica.

On land, vegetation steadily disappears, giving way to man-made deserts of concrete and asphalt—cities, industrial areas, airports, shopping centers—that do not produce any oxygen at all. It is possible that the activities of land plants in producing oxygen are chiefly transient, and that they therefore play a role in the oxygen cycle different from the green algae. In any event, the continued destruction of marine and terrestrial vegetation all over the world may lead to a sudden collapse of the oxygen balance of the biosphere. Such a dangerous situation threatens all living organisms of our planet. We still do not know enough about the rates of photosynthesis and oxygen consumption to be able to predict when the oxygen balance of the air, water, and soil will be seriously upset by the accelerating elimination of green plants. What we do know is that we are blindly approaching the critical point of global vegetation destruction that could wipe out the possibility of life on earth.

These threats to the world's supply of oxygen have been much debated in scientific circles during the last years. As far as I know, only five scientists have so far (in 1970) suggested that there are no reasons to believe in an eco-catastrophe due to the decline of the oxygen supply in the atmosphere and the oceans in the fore-

* This proportion, given by Cole (1968), might be an overestimation. Other recent estimates of the global production of organic matter indicate that the oceans contribute less than one-third to the world total (Ryther 1970, Whitaker 1970). However, even this proportion is important, particularly when seen in the light of the rapid disappearance of the world's forest cover.

seeable future. Molecular oxygen is said to be one resource that is virtually unlimited.

It remains to be seen what will happen to "resource air." Until very recent times, man thought that other renewable natural resources were also inexhaustible, but today we know that this was a serious mistake.

Moreover, also the functioning of the cycle by which atmospheric nitrogen is made available is threatened. Nitrogen is another necessity to allow life to exist on the earth. This cycle depends on a wide variety of organisms, including soil bacteria and fungi, which on an increasing scale for the last thirty years are being destroyed by persistent chemicals accumulated in the soil.

The oxygen cycle may also be upset by the burning—i.e., the very rapid oxidation—of fossil fuels such as coal, oil, and natural gas. The oxygen consumed by the combustion of these fuels is replenished by photosynthesis—mainly, as mentioned above, in the oceans, but also in swamps, bogs, and certain sediments, both freshwater and marine. However, in burning fossil fuels in increasingly larger quantities man has raised the carbon dioxide content of the atmosphere about 10 percent over the past hundred years. Although the oceans have a great potential for absorbing carbon dioxide, they cannot under present conditions prevent an increased tension in the air. Tropical rain forests can assimilate considerable amounts of carbon dioxide, but this is only a delaying action. Moreover, the tropical rain forests are tending to disappear as human activities advance.

An effect of the increasing concentration of atmospheric carbon dioxide is a rise in the earth's temperature due to the so-called greenhouse effect: carbon dioxide allows sunlight to pass through but reduces reradiation of heat into the atmosphere. If we continue to burn fossil fuels at an accelerating rate, the increase in retained heat could lead to a slow melting of the polar ice caps. This in turn, it is estimated, could raise the sea level four hundred feet and inundate vast land areas, including all of the world's present major ports.

Observations for 1958 and 1970 show that the carbon dioxide concentration in the atmosphere increased from 312 to 320 parts per million. This is an average annual increase of 0.7 parts per million. If this rate continues, it would double man-made carbon dioxide accumulations in the atmosphere in about twenty-three years. However, for the moment the earth's surface temperature is falling rather than rising. This means that the heating effects of carbon dioxide are neutralized by other factors, which may either be natural or man-made.

We do not know yet whether recent climatic changes are due to natural phenomena or partly also to man's activities, but in the history of earth major and minor climatic fluctuations are normal processes and certainly they still are. On the other hand, atmospheric man-made pollution has increased markedly in a measurable way and cannot be neglected as an important factor, though no proofs exist that urban or industrial pollution is the prime cause of the recent cooling trend. It has been suggested that a major part of the recent temperature variation is due to particles introduced to the upper atmosphere by volcanic eruptions.

Man can change atmospheric temperatures slightly, however, and thus significantly affect climate in at least seven ways which were listed in 1970 in the First Annual Report to the U. S. Congress by the Council on Environmental Quality:

1. By increasing the carbon dioxide content of the atmosphere by burning fossil fuels.
2. By decreasing atmospheric transparency with aerosols (tiny solid or liquid particles floating in the air) from industry, automobiles, and home heating units.
3. By decreasing atmospheric transparency by introducing dust (particles larger than aerosols), primarily through improper agricultural practices.
4. By altering the thermal properties of the stratosphere (upper atmosphere) with water from the combustion of jet fuels.
5. By heating the atmosphere by burning fossil and nuclear fuels.
6. By changing the ability of the earth's surface to reflect sun radiation back into space through urbanization, agriculture, deforestation, and reservoirs.
7. By altering the rate of transfer of thermal energy and momentum between the oceans and atmosphere by spilling oil on the water's surface. Such oil films on the ocean come from incomplete combustion or spills from sources such as vessels and drilling towers.

Number 4 will undoubtedly have attention focussed on it when future supersonic jets will travel in the stratosphere, because altering the water content of the upper atmosphere can upset the earth's radiation balance. In and above the stratosphere water vapor stays for about eighteen months, on an average, because it mixes slowly with the lower atmosphere. Intense supersonic jet traffic in the stratosphere could increase the water vapor content up to 10 percent globally and 60 percent locally and augment natural fluctuations of climate.

Man is inadvertently playing with physical forces of a magnitude he cannot master without worldwide intensified research on the long-term effects of artificial atmospheric alterations.

Air Pollution

The far-reaching poisoning of the air has not been known very long. It has recently struck urban and industrial communities of North America, Europe, and Asia as a result of heavily increased discharges of chemical waste in the air. It must be realized that this situation will prevail and gradually become worse unless effective controls are undertaken. The potential danger is always present over certain areas, though "pollution eruptions" with strong poisoning effects on the air we breathe are often determined, as mentioned above, by occasional, local atmospheric conditions.

The most serious air pollution catastrophe up to the time this book went to press occurred in London in December, 1952. As a direct consequence of smoke from domestic fires during a fog period of five days, a poisonous smog accumulated over the city that caused the death of over 4,000 people. No age groups were spared, but it chiefly struck older persons and children less than a year old. During the worst days of the smog the human mortality was equal to that of the cholera epidemic in London in 1866. Moreover, many thousands of people suffered a worsening of such illnesses as bronchitis and heart disease.

Similar air pollution catastrophes with fewer casualties have struck Los Angeles, where smog killed 300 people during a three-week period in 1954; the Meuse Valley in Belgium, where 60 people died and several thousand were seriously poisoned during a series of smog days in 1930; in Poza Rica, Mexico, where 28 died and 320 were poisoned by an accidental release of airborne sulphurous gases in 1950; and at Donora near Pittsburgh, Pennsylvania, where smog caused the death of 20 persons and poisoned nearly 7,000—50 percent of the population—during four days of 1948.

Obviously, the lethal dangers from air pollution have been chiefly observed and discussed in connection with such disasters, which are spectacular manifestations of the underlying condition. It is the mass, long-term effects of air pollution that constitute the most serious threat to living organisms. The slow, daily, invisible poisoning gradually breaks down human health and lowers the resistance to other diseases. It is calculated that about 10,000 people die annually in the United States due to breathing what is presently considered normal air in the major cities.

Although Los Angeles has the strictest smog control laws in the world, there is no progress against smog because the population is growing too fast. In 1970 New York and Tokyo were exposed to serious smogs which greatly affected life and community functions of these two large cities. Also, the atmospheric pollution at Fairbanks, Alaska, is as bad as anywhere in industrial areas, and the concentration of lead in the air during extreme conditions probably exceeds the values measured in most urban areas. Local conditions create ice fogs ten meters high (thirty meters over the city center), and visibility is reduced to less than ten meters, recalling the heyday of the old "London particular." The situation is particularly acute when the air temperature remains below −40°C for a week or more.

Los Angeles uses very few of the fuels primarily responsible for the smoke problems of cities elsewhere. This situation pinpoints the principal sources of the Los Angeles smog: hydrocarbons and nitrogen oxides from automobile exhausts. This means that other major cities also are afflicted by photochemical smog coupled with other forms of air pollution.

Aircraft also pollute the air heavily. The takeoff of one Boeing 707 has the same pollution effect as 6850 accelerating Volkswagens. Near the airport of Munich in Germany 20 hectares of pines are dying because of airplane exhausts.

In fact, in the United States, transportation, particularly automobiles, is the greatest source of air pollution—accounting for 42 percent of all air pollutants by weight. One study estimates that, in cities, exhausts from motor vehicles constitute 75 percent, by weight, of all urban smog.

In New York some streets in Lower Manhattan were in 1970 declared car-free on weekends. The carbon-dioxide count on some autoless streets fell by as much as 90 percent, while the average noise level shrank from 78 decibels (shouting level) to 58 decibels (conversation level). Smog-shrouded Tokyo decided at the end of 1970 to ban all cars from no fewer than 122 of the city's busiest streets on Sundays. The carbon-monoxide level dropped as much as 80 percent.

In December, 1970, the U. S. Congress passed a tough clean-air bill setting uniform standards for industrial plants and ordering a 90 percent reduction in auto emissions by 1975. This important measure will undoubtedly force technological research to find solutions against industrial and automobile exhaust pollution which are a necessity for urban populations.

Air pollution has other consequences. Weather and climate change, light intensity lessens, water and soils are affected negatively, plant and animal populations, agricultural crops, and forests are damaged; the properties of a whole gamut of materials—metal, textile, leather, paper, rubber—deteriorate.

That visibility is reduced by air pollution over cities and industrial areas has certainly been observed by every city dweller. In one year alone, low visibility from smoke, haze, and dust was the suspected cause of fifteen to twenty aircraft crashes. During the smog over the eastern United States in July, 1970, visibility was almost totally obscured in some areas and interfered in the safe operation of planes and vehicles.

What air pollution costs in damages alone cannot be precisely calculated. In addition, the economic loss attributable to air pollution is probably not yet fully realized, because it damages almost everything in an invisible way. In the United States the economy is suffering $10–12 billion a year because of air pollution, according to Senator Gaylord Nelson of Wisconsin. The U. S. Council of Environmental Quality has, in its First Annual Report (1970) to the Congress, concluded that these costs amount to many billions of dollars a year. Economic studies are beginning to identify some of the more obvious costs. To paint steel structures damaged by air pollution runs an estimated $100 million a year. Commercial laundering, cleaning, and dyeing of fabrics soiled by air pollution cost about $800 million. Washing cars dirtied by air pollution costs about $240 million. Damage to agricultural crops and livestock is put at $500 million a year or more. Adverse effects of air pollution on air travel cost from $40 to $80 million a year. Even more difficult to tie down are the costs of replacing and protecting precision instruments or maintaining cleanliness in the production of foods, beverages, and other consumables. It is equally difficult to assess damage, soiling, and added maintenance to homes and furnishings or how air pollution acts on property values. The cost of fuels wasted in incomplete combustion and of valuable and potentially recoverable resources such as sulphur wasted into the air is also hard to count. It is still more difficult to determine the dollar value of medical costs and time lost from work because of air pollution—or to calculate the resulting fall in productivity of business and industry.

Radioactivity, radiation, and nuclear fallout are briefly discussed in Chapter 8, Man.

Noise Pollution

Another kind of airborne pollution is artificial noise. In modern cities, where millions of people are in permanent residence, the multiple and insidious ill effects of noise on mental and physical health is obvious but still inadequately recognized. Noise pollution also occurs outside of the cities. Sonic booms are perhaps the most threatening of the technological noises. The effects of sonic booms from commercial supersonic transports will certainly be harmful, but we do not know their full implications.

Noise is difficult to define. What to some people is acoustically beautiful or entertaining is to others undesirable, even unbearable. Individual reactions also depend on the circumstances. Hence, the tolerance threshold expressed in decibels is almost impossible to determine. However, four general categories of effects from noise pollution may be distinguished: annoyance, disruption of activity, loss of hearing, physical and mental deterioration. Shocks from sonic booms may include effects that comprise a fifth, still unknown, category.

So far the most severe noise pollution consequences are generally to be found in the industrial environment. Excessive exposure to mechanical noise for long periods may cause irreversible hearing loss. In the United States it is estimated that up to 16 million American workers are threatened with hearing damage.

Legislation concerning unwanted noise and noise control has been introduced or is under preparation in several countries. Other governments are studying the problem. Noise pollution is undoubtedly here to stay—a regrettable characteristic nuisance of our machine civilization.

Effects on Plants

Plants react significantly to air pollution sooner than any other organisms. Lichens and mosses are particularly sensitive and may be used as living indicators of air pollution. In many urban areas near heavy motor vehicle traffic, lichens growing on trees have been exterminated by poisonous gases. This is a direct killing effect, but other effects may be even more serious. Air pollution is detrimental to the environmental metabolic cycle of plants. It inhibits the photosynthetic process whereby they take in carbon dioxide and give off oxygen. Ultraviolet light is often excluded by smoke and smog, with detrimental effect on all living organisms. Further, the lessening of light intensity by air pollution may result in slower growth. True, the environmental change caused by reduced light energy may favour other plant species more suited and adaptable to the changed level of intensity, but the importance of direct lethal effects on plants and plant communities exposed to air pollution many times surpasses the few advantages. The damage caused by air pollution to agricultural crops in Great Britain in 1954 was estimated at about £10 million a year. In the United States the annual damage to tobacco and citrus crops is calculated at about $60 million. In Czechoslovakia damage to forests has been estimated to be $40 million a year.

In and around cities many species of trees slowly die due to air pollution. This is, for example, the case of pines (*Pinus sylvestris*) around Helsinki. Smog from Los Angeles is affecting trees even in the San Bernardino National Forest sixty miles away from the city. Smog injury showed up in 100,000 acres of this National Forest, and taking private land into account also, no fewer than 1.3 million trees are estimated to be blighted. About half of mature ponderosa and Jeffrey pines show some damage, 15 percent being severely affected and 3 percent are already dead. Basic cause is believed to be destruction of chlorophyll by the high ozone content of the smog.

Sulphur dioxide fumes from a copper smelting plant in Copper Basin, Tennessee, damaged 30,000 acres of timberland during the nineteenth century. Much of this former forest land is still barren. Ferroride and sulphur oxides in Florida have damaged large numbers of pines and citrus orchards. Also in California citrus growths slowly decline due to the Los Angeles smog. In New Jersey, pollution injury to vegetation has been recorded in every county.

Effects on Animals

The deleterious effects of air pollution on animals are chiefly known in cattle and sheep, which have been seriously harmed by grazing in contaminated pastures. This is difficult to foresee and prevent because the plants eaten may not necessarily show visible signs of being poisoned. Not until the physical quality of livestock has been markedly lowered is the situation manifest. It may be that a comparable condition exists relative to human beings eating vegetables.

A spectacular case of air pollutant harm to animals is that of some cattle that were brought to London for an agricultural exhibition during the smog period of 1952. These animals had to be destroyed. Autopsies showed they were afflicted with heart disease and respiratory ailments. Apparently air pollution affects the capillary respiratory passages in mammals and probably in other terrestrial vertebrates as well. A new livestock disease, fluorosis, is directly linked to air pollution. It develops a crippling condition in livestock grazing on fluoride-tainted vegetation.

Air pollution through aerial pesticide spraying of water, crops, and forests also causes casualties in domestic and wild animals. Such pollution is discussed in the chapter on Vegetation.

Obviously all forms of life, from flying creatures to microorganisms living in soil and water, are adversely affected by pollution of the air, but the deteriorating processes are slow, and our knowledge of these changes is still grossly inadequate.

Effects on Man

Since man is the numerically dominant vertebrate species in urban biocommunities, he is the chief victim of the air pollution chain. Some of the spectacular air pollution disasters that in a short time eliminated large numbers of human beings or cut down their potential life expectancy have been mentioned. It is now established that air pollution contributes to the incidence of clinical diseases like emphysema, bronchitis, and other respiratory ailments. In the United States emphysema has doubled in frequency every five years. It is significant that bronchitis is more prevalent in industrial countries than in agrarian ones. In Great Britain chronic bronchitis is the leading mortality factor among human males above forty-five years of age. The death frequency is twenty times higher than in less industrialized Sweden.

Polluted air also causes higher mortality rates for other defects such as cancer and arteriosclerotic heart disease. Lung cancer has increased tremendously during the recent decades. In fact, no other deadly human disease has increased to the same extent. In the United States lung cancer was previously a rarity, but its incidence has increased forty times during fifty years. From 1930 to 1951 the number of deaths due to lung cancer increased by 50 percent. Only a fraction of this increase in mortality can be attributed to the increasing mean age. At present lung cancer kills more persons in the United States than all other cancer types combined. A similar lung cancer mortality trend is pronounced in Great Britain and a number of other European countries. Everywhere the frequency of lung cancer in city-dwelling populations is higher than for those living in the country. People emigrating from Great Britain to South Africa and New Zealand show a lower frequency of lung cancer than people remaining in the British Isles, but higher than that of Africans and Maori. The lungs of countryside inhabitants have the normal pale pink color, while those of people living in towns are usually greyish.

An adult human requires more than thirty pounds of air a day. He breathes about twenty times each minute, each time inhaling six pints. If he lives in an area that has a dirty atmosphere, he cannot avoid a continuously repeated filling of his lungs with dirty air. When carbon monoxide—a colorless and odorless poisonous gas—is inhaled it displaces oxygen in the blood and reduces the amount carried to the body tissues. The carbon monoxide is the most common air pollutant in areas with heavy automobile traffic. In the United States more than 100 million tons of carbon monoxide were emitted in 1968, of which an overwhelming bulk came from gasoline-powered vehicles. When inhaled at pollution levels commonly occurring in cities, carbon monoxide can slow the reactions of even the healthiest persons and constitutes an increasing risk for accidents.

What to Do?

Since everyone has to breathe, air pollution has in recent years—but much too late—aroused the attention and concern of governments. At long last it has become a consequential political issue. As yet only a few legislative measures have been taken in this complex field. The United States, for example, recently passed its first federal law for control of air pollution. The full-scale legislative action yet to come must be based on intensified technical, medical, climatological, meteorological, ecological, and biological research, and on international cooperation. Continent-wide or, even better, global conventions are required to enact basic air quality regulations that every nation should be persuaded to implement.

It is interesting to note that recently there are indications in the United States that environmental antipollution control has become big business. In 1969, for example, $94 million were spent on the purchase of air pollution control devices. This is an industrial antidote to the pollution.

Cleaner air is certainly one of the most urgent conservation necessities facing our world. However, as of now only a relatively small area of the globe is affected by constant air pollution. Man still has the opportunity to solve a critical problem by eliminating a barrier to life and environmental health before very large areas and populations are afflicted. This undertaking will be very costly and will require much technical skill. It is imperative that governments and industries take action without delay.

CHAPTER THREE

The Sea

The sea is one of the world's few remaining wilderness areas. Man has not yet been able to dominate marine habitats, nor has he succeeded in utilizing the full potential of marine resources. The world harvest of marine fish in 1970 was about 57 million tons. In addition, invertebrates and algae increase this total of marine cropping. Even if this production were doubled, it would not be enough to feed the hungry world population of today. While the biological productivity of the sea is enormous, we do not know how much man can take out of it. Will the sea save man's future? Can the marine resources yield be increased to where it will meet the food requirements in protein-deficient areas? Any answer to these questions would be a speculation. Though many marine resources have been heavily overexploited and even exterminated, others are untapped because we do not know technically how to utilize them. Obviously the various life zones of the sea constitute natural resources of great value and should be conserved and managed wisely. This is a matter of international concern, for the seas have no national boundaries. Moreover, they are all interconnected, without physical barriers—except for such factors as temperature and salinity—so that the organisms living in the sea can be thought of as forming global populations. The marine resources of international waters belong to nobody or to everybody—not to a few countries alone. Yet three nations are heedlessly wiping out one of the most important marine resources of our globe: the whale.

Ecology and Sea Life Cycles

Plants and animals living in the sea depend on a wide spectrum of environmental factors just as do terrestrial organisms. Marine life responds to waves, tides, currents, temperatures, light intensity, salinity, oxygen supply, chemical cycles, depth, pressure, types of bottom, and, of course, the living world of other plants and animals in the marine environment.

The marine life cycle is the same in principle as in freshwater and on land: excreta and decaying remains of dead plants and animals are broken down by bacteria and release organic compounds in the form of phosphates and nitrates that are absorbed by microscopic living organisms such as diatoms and flagellates. The organisms in turn build up tissues that serve as food for other living beings, and the food chain continues. Although the interrelationships of marine organisms are infinitely varied and complex, the life cycles are basically the same: the whole life fertility of the seas dependent on microscopic algae that depend in turn on an orderly breakdown of dead plants and animals.

In the sea are many food chains, the links consisting of vast varieties of plankton, mollusks, arthropods, fish, birds, and marine reptiles and mammals. Plankton is the mass of small animal and plant life that drifts with the currents and tides of the sea—small animals such as protozoans, larval fish, and crustaceans, some larger ones such as jellyfish, and tiny plants such as algae. Plankton organisms dominate the sea populations in number and are the base of almost all the food chains. In addition, many fish and mammals utilize plankton directly. For example, two sharks that are the largest fish in the world and the baleen whales, the largest mammals that have ever existed, feed directly on plankton. Man likewise utilizes plankton as food, as transferred through other groups of marine animals. However, the energy loss through the different links of food chains is considerable, which is clearly demonstrated by the baleen whales, that build up their enormous bodies by feeding directly on living plankton organisms.

Although the sea appears to be ecologically rather uniform, it actually offers a wide range of habitats, chemical and physical. Life zones exist in practically all parts of the sea, from deep bottoms to the surface. Various types of bottoms make for as many different kinds of habitats. Rocks, stones, gravel, sand, mud, shells, and corals constitute different environments, each with a distinctive fauna varying with depths, salinity, tidal rhythms, and so on. Most marine organisms float or swim in the pelagic zone, where the greatest abundance of life forms is found.

Many marine habitats are vulnerable and can easily be destroyed. This is particularly true of reefs that are formed by the accumulation of calcareous skeletons of myriads of tiny animals. Coral reefs belong to one of the most productive ecosystems in the world. They have a special fauna extremely rich in fish species. Many pelagic fish reproduce in the coral reef zone and/or spend their early life stages there. The importance of coral reefs as an underwater resource for public recreational use, snorkeling and skindiving, is rapidly increasing.

Disastrous effects on marine life zones are continuously caused by oil discharges from ships in the open sea. Like the extermination of whales, this form of modern piracy results from the unfortunate situation that international waters are considered to be a "no man's sea," where neither nations nor individuals have responsibility.

Productivity

The sea covers almost three-quarters of the earth's surface. However, biomass and productivity vary greatly within this enormous ecosystem. Man is getting only one-thirtieth as much produce from marine resources as from terrestrial ones. On the other hand, fish and crustaceans account for almost one-fifth of the world's consumption of animal egg white proteins. The potential of fish as a food resource, relative to what agriculture can produce in many countries, is undoubtedly underestimated.

The most productive areas of the sea are those where nutritive mineral elements give optimal growth conditions to planktonic plants. Fortunately for man, these highly productive regions are in many cases located rather close to the continents. Off Peru, for example, one of the world's greatest fishing resources is the Peruvian anchovy (*Engraulis ringens*). It amounts to almost 18 percent of the total world fish production, yielding an annual average catch of about 9 million tons. Moreover, the Peruvian anchovy is also essential to the economically important guano—bird manure—industry in Peru, since it is the staple food of about 35 million cormorants, boobies, and pelicans that breed on the Chincha Islands off Peru. These birds produce guano in tremendous quantities. Sold for fertilizer, its economic yield to Peru is enormous, competing with the income from the anchovy fisheries. These two resources, fish and fertilizer, so essential for the Peruvian economy, are based entirely on the existence of tiny plankton organisms in the cold Humboldt Current flowing in the Pacific along a part of South America's coast.

Other highly productive parts of the oceans are coastal waters of the northern Pacific, the North Atlantic, and the Arctic Ocean, as well as the North Sea and an area of the Atlantic off Portugal. An extremely rich productivity of microorganisms has developed in the Antarctic Ocean. The plankton there were supporting the greatest concentration of baleen whales in the world, a resource that man has unnecessarily depleted to a fraction of its former abundance. In general, polar and temperate oceans are more productive than subtropical and tropical ones.

We still do not know enough about ocean productivity. Since evaluations and estimates have been based on fishery and whaling catches, they relate only to particular species, excluding the majority of marine life forms. The seas are vast pastures, but the grazing efficacy of the vertebrates on phytoplankton (plants) and zooplankton (animals) is little known. Moreover, the distribution of zooplankton is patchy, and their quality varies greatly. In the sea only a few species of economically important fish eat plants directly. Copepods form the main link in the food chain between marine plants and fish. They are small planktonic crustaceans occurring in all the oceans and many of them graze phytoplankton. In coastal waters planktonic stages of bottom-living worms, echinoderms such as starfish, and mollusks feed on phytoplankton. Thus zooplankton grazing on phytoplankton have a key position in the food chain of economically important marine species.

The importance of adequate food densities emphasizes the vulnerability of marine ecosystems and their productivity. If just one link in the food chain of life cycle within a region is damaged, the whole community will suffer. If one main link is temporarily eliminated, the whole ecosystem will collapse, and the productivity will be nil or heavily reduced for an ensuing period of time. The population sizes of such important fish as herrings, sardines, and anchovies may be influenced negatively for decades. Human exploitation of ocean resources must therefore be intelligent, based on a full knowledge of their productivity, life cycles, population shifts, and migrations (both vertical and horizontal) so that the annual harvest is synchronized with all these phenomena without diminishing the capital by reducing the breeding rate—in other words, so that sustained yield will be obtained. This means, for example, that one or several countries should not exploit in different seasons of the year the same migratory populations of marine animals in two or more regions without being aware that in reality only one stock is involved.

Intensified research can avoid such mistakes, and consequent conservation measures will pay their way. Full use of a marine environment can be achieved only if it is wisely managed and conserved, exactly as with a terrestrial environment. Managing marine communities is not simple because the populations are in no way stable. Flux is constant and dynamic changes are the norm.

Overfishing

As long as the essential environmental conditions of life, including the chief links in the food chain, exist for a species, man's reasonable harvest of it will have little or no effect on the population. If intense fishing diminishes or even exterminates a local population, it will regenerate providing the habitat has not been upset and the fishing intensity is regulated.

Many species of fish show long-term fluctuations in number as a response to various environmental factors. This is often the case with various species of herrings, sardines, and anchovies, generally having nothing to do with human activities.

Steady population decline of a marine species usually reflects overexploitation. Decrease in abundance often leads to more intensive fishing methods, for instance, going to finer-meshed trawls that catch smaller sizes of the declining species. The proportion of immature fish in the catch increases, with a direct negative effect on the reproduction of the species. The population dwindles and so does the mean size of the fish caught. On the other hand, reductions of local populations through fishing may have a positive effect on the mean size of the species involved because the intraspecific competition decreases and more food becomes available for each individual.

Overfishing has occurred and is still going on in many sea regions. Evidence of decline, exhaustion, and even local extermination of economically important fish and invertebrate populations is so numerous that examples from all over the world can be selected at random. Population declines of anadromous fish—species that migrate from freshwater to the sea or vice versa, like salmons—are excluded here because their decrease has until recent time been due chiefly to factors outside the marine habitats. Statistical records from the western North Atlantic indicate that the catches, for example, of cod (*Gadus morhua*), yellow-tailed flounder (*Limanda ferruginea*), and halibut (*Hippoglossus hippoglossus*) are far below those of the past. The halibut was once landed to the extent of 14 million pounds per year, but, as a result of overfishing, the quantity gradually decreased to about 0.2 million pounds a year, the annual catches of the United States in 1961–1965.

The same species of halibut lives also in the Pacific, where it is even more important to the North American fisheries than the Atlantic population. In the United States the ratio of the Pacific to Atlantic halibut catch is about 40 to 1. The fishing of Pacific halibut at the start of the century produced annual catches between 50 and 60 million pounds, the peak coming in 1915 with a catch of 63,254,000 pounds. The continuously high harvest was achieved by intensified fishing and a gradually widened range of operations. When the limits of geographic expansion were reached, the annual catches began to decrease, despite increased fishing. The catch per unit of gear declined at an alarming rate. In 1906 the average catch per unit was 250 pounds, but in 1930 it had dropped to 35 pounds. It became evident that the Pacific halibut fishing industry had for years misused the capital by heavily overfishing the population. In the 1920's large scale halibut research was undertaken in the northern Pacific. The findings led to the adoption of regulatory measures through a treaty between Canada and the United States. This conserving action gave the Pacific halibut a chance to regenerate, bringing remarkable changes in the halibut situation. The life-span of the halibut is about twenty-five years, and the species may reach an age of fifty years. Within less than twenty years after the conservation measures were applied, the abundance of the Pacific halibut was twice what it had been in 1931, and the annual catches increased as much as 15 million pounds, taken with one-third less fishing effort. Since then the rehabilitation of the Pacific halibut has continued. In the 1960's the annual catches have oscillated between 26 and 40 million pounds for the United States and between 29 and 37 million pounds for Canada.

The Pacific halibut story is probably the best example of successful management of a marine resource. It shows clearly how wise cooperation with nature based on ecological knowledge gives a stabilized productivity at a high level and with immediate profits —more fish with less fishing.

In other marine regions, formerly rich fisheries have disappeared entirely due to overexploitation. A tragic example is that of the waters around the Juan Fernandez Islands off Chile, where an abundance of economically important fish and crustaceans, particularly lobsters, was once the basis for the island economy. These resources were exploited too heavily, with no regulation. The lobsters have long since vanished and so have many fish species. The populations of the remaining fish are so thinned out that the fisheries are no longer profitable. In the North Sea—so important for international fisheries—most of the commercial exploited fish populations of various species are kept down by overexploitation for prolonged periods.

Local exterminations of fish species have been reported from other fishing sea areas. Since spearfishing by skindivers has become popular, it is claimed that some species of fish, and even whole fish populations, have disappeared from intensively frequented coasts and islands in the Caribbean and Mediterranean seas and in the Indian Ocean. In this case it is recreational fishing that menaces certain fish, especially those species that have a very narrow ecological niche and therefore a restricted distribution. "Dynamite fishing" along rocky coasts and coral reefs kills blindly and ecologically and is an extremely dangerous method leading to depletion if frequently used.

Despite unwise fishing in all seas, man has not yet exterminated any marine fishes, since they are widespread and generally prolific. However, marine fish populations will be depleted sooner or later if the fishing pressure is not regulated with guidance from ecologic research and international agreements. Bottom fish are particularly vulnerable, while pelagic species seem to be more resistant to human predation.

Marine fish and invertebrates represent high-quality protein. The sea provides at present almost one-fifth of the animal protein consumed by human beings. This important marine food supply must be managed properly in the interest of mankind, so the productivity of the sea does not diminish through shortsighted exploitation by a few countries.

Pollution

Man has created imbalance in marine environments by other means than overexploitation. The pollution of oceans and seas and the adverse effects on marine living resources have reached alarming proportions all over the world, even in remote and relatively little frequented areas like the Arctic and Antarctic Oceans. The Baltic Sea is probably the most polluted marine water in the world, the North Sea fisheries are already affected negatively and parts of the Mediterranean Sea show high and hazardous levels of pollution.

Discharges of oil in order to clean the tanks of oil tankers, and grounding of tankers with subsequent releases of large quantities of crude oil are disasters that heavily affect marine populations. The best-known disaster was the *Torrey Canyon,* an oil tanker of considerable size that ran aground on the Seven Stones Reef off the coast of Cornwall, England, in 1967. Much of the tanker's oil was lost in the sea, creating an oil slick hundreds of miles in length and locally over a foot thick. It polluted not only the sea but also many areas of the coast. Many boulders of the upper part of the beaches were almost completely hidden by oil.

The effect of this disaster on marine life, particularly seabirds, was very serious. In the efforts to ameliorate the consequences, man made a serious mistake. Intervening without consulting with ecologists, the British authorities made things even worse by spraying large quantities of toxic chemical detergents and emulsifying agents on the oil. The mixture of emulsified oil and detergent was far more toxic than the oil itself, at least in coastal waters. All or nearly all detergents are toxic themselves, but once the oil is emulsified it ceases to float as a surface layer and becomes instead an actual part of the pelagic environment, where it may be taken into the gills of fish or ingested by filter-feeding organisms. When oil slicks are sprayed with detergents, pelagic fish may be killed. In addition, since much of the emulsified mixture is likely to remain near the surface, it may do considerable harm to plankton, which ascend to the surface at night. British studies of the *Torrey Canyon* disaster showed that while oil killed 30 percent of planktonic organisms, detergents destroyed up to 96 percent. Sinking oil probably accumulates on the bottom, which may be dangerous for organisms living there.

What is known with certainty is the terrible effect on birds of oil discharges in the sea. At sea, flying flocks of ducks and auks often swoop down on the patches of oil, where the deep-sea roll is less heavy. These oil patches serve as veritable deathtraps. The oil-soaked birds can neither fly nor dive, and their plumage no longer protects them against cold and wet. If the oil covers an area of the plumage no larger than the size of a coin and the water is winter-cold, it can lower the bird's temperature enough to cause its death. Thus the oil-smeared birds either starve or freeze to death, or become so wet and heavy that they drown.

Almost every year tens of thousands of diving ducks and auks are destroyed by oil discharged in the Baltic and North seas alone. In the winter of 1952 about 30,000 ducks, chiefly the long-tailed duck (*Clangula hyemalis*), died off the Baltic island of Gotland. Six years later a similar catastrophe struck the same area, within a few days killing another 30,000 oil-soaked ducks. The grounding of a German ship off the estuary of the Elbe in 1955 caused a release of 6,000 tons of oil that killed about 100,000 seabirds. Large numbers of shorebirds are also destroyed when oil is driven in over shallow coastal waters. Aside from the cruelty, such tragic destruction is a serious economic waste of a natural resource utilized in many countries. To these losses must be added the damage to other living organisms and human property, and to ruined recreational opportunities along destroyed beaches.

It has been calculated that until recent time tanker fleets have discharged annually while cleaning their tanks between one and two million tons of oil into the sea. Fortunately this volume has decreased. The major oil companies now use the so-called load-on-top system, which has considerably reduced the quantity discharged. However, many tankers and other ships still discharge crude oil, cases of pollution are frequent, and the destruction of seabirds and other marine resources continues even within protected zones. At present the Convention for the Prevention of the Pollution of the Sea by Waste Oil stipulates that oil discharges may not be made in designated sea regions, chiefly located off the coasts, but it is difficult to trace an offending ship. What is needed is a total ban on oil discharges into the sea.

Oil can pollute the ocean in other ways as well. Drilling for oil in the sea is a potential damage risk. Although drillings in the Gulf of Mexico off Louisiana and Texas had not led to any big leaks from 1957 until 1970, even during hurricanes, geological conditions in other coastal areas may be less stable—particularly in earthquake-prone regions. The undersea oil leak catastrophe in the Santa Barbara Channel of California in January and February of 1969 was a dire warning. An estimated 21,000 gallons of crude oil per day poured during twelve days from the ocean floor with a massive infusion of heavy drilling mud. The oil slick covered hundreds of square miles in the ocean channel and was washed up along more than forty miles of shoreline.

The long-term effects on marine life of the Santa Barbara disaster are hard to foresee. Disturbances of food chains resulting from killing of plankton and other organisms may take years to repair. At the end of July, 1969, nearly all beaches in the Santa Barbara vicinity showed oil stains near the high tide line. Oil could be found along approximately forty miles of coastline including a sanctuary. Two widespread oil slicks were still seen west of Santa Barbara and south of Carpenteria as late as July 27.

In 1970 the President of the United States proposed to the Congress that the Federal Government cancel twenty federal leases along the Santa Barbara Channel. Obviously, the Federal Government should never have permitted oil drilling in the channel where there are pronounced recreational, fishing, commercial, ecological and esthetic values represented.

Unfortunately, the Santa Barbara oil bill was not accepted by Congress, but in May, 1971, the Department of the Interior submitted new legislation to Congress to terminate thirty-five federal oil and gas leases in the Santa Barbara Channel, and to create a National Energy Reserve in the channel.

In February, 1970, crude oil from exploited undersea wells, ten miles off the shore of Louisiana, spurted out and floated turgidly on the clear waters of the Gulf of Mexico. It ignited at the surface. As long as the flames blazed, the oil was consumed. When the fire was extinguished, the oil wells continued spewing 1,000 gallons of spill a day, producing a thick belt of oil, one mile wide and seventeen miles long, which floated toward the main oyster beds off Louisiana which are so important for the fishery industry there. Desperate efforts were made to stop the disaster. The leaking wells belonged to the Chevron Oil Company, that had not followed what federal regulations required: storm chokes, devices which should have been installed 1,000 feet below the surface to shut off the oil automatically when the flow became excessive. Chevron admitted that 120 of its 292 offshore wells did not have storm chokes, even though federal inspectors had previously found the devices in place. Removing of chokes increases production from an oil well. . . .

These incidents off California and Louisiana clearly demonstrate the potential risks for disastrous pollution of coastal waters by oil spills from marine wells.

But not just oil is a source of pollution to the sea. Coastal waters, particularly near large cities or industrial installations, are simply used as waste dumps. In 1969 off the U. S. coasts alone 37 million tons of solid waste were dumped. Waste disposal offshore has a negative influence on many marine organisms and may in the long run threaten not only fishery activities in the area but may also act detrimentally on a much larger scale by reducing the production of oxygen. Also, enormous quantities of toxic chemicals reach the sea annually by way of rivers entering the sea and fallout from the winds.

A drastic example of how the lethal methyl mercury can poison fish living in the open sea is the tuna. This fish represents the last link in a marine food chain. In December, 1970, it was found in the United States that millions of cans of tuna contained dangerous levels of mercury. The U. S. Food and Drug Administration confirmed that as much as 23 percent of the tuna fish being sold across the United States might be contaminated with poisonous mercury. There are 900 million cans of tuna packed in the United States each year. The poisoned tuna came from all eight major commercial fishing areas off the United States, where even higher levels of mercury were being found in some swordfish. Also on the European side of the Atlantic tuna fished by British fishermen contained similar quantities (0.37–1.12 parts per million) as in the United States.

In May, 1971, it was shown, after a series of tests by the U. S. Food and Drug Administration, that 95 percent of the swordfish samples contained dangerous amounts of mercury. Almost all samples analyzed had more than the allowable limit of 0.5 part per million. The average level of contamination was twice as high, and 8 percent of the samples actually tested out at 1.5 parts per million. Of 853 samples examined, only 42 passed the safety level. This discovery led to a withdrawal of nearly 5 million pounds of swordfish from the market and, probably, the end of swordfish consumption in the United States where it has been a popular and widespread food.

The evidence obtained from tuna and swordfish shows how even mercury compounds are spread in the open sea, presumably originating from industrial wastes washed into the sea. If predators such as the tuna and the swordfish are contaminated by mercury, it is certain that many other marine fish are also contaminated.

Mercury in fish has already caused lethal effects in man. In 1953 and 1961 inhabitants at Maranata in Japan ate marine fish that had taken up mercury from a chemical plant's discharge in a river. Forty-nine persons died and nineteen babies with various congenital defects were born to mothers who had eaten the polluted fish. The occurrence also of DDT and its derivatives in the open ocean is recorded by the damaging concentrations in the systems of animals of the sea. While this form of poisoning is a grave hazard to the ocean plankton populations, organisms in estuaries and coastal waters are in the greatest danger because they are more directly exposed to the steady flood of persistent toxic chemicals. The breeding potential of populations of estuarine and coastal fish and invertebrates is threatened, which is in turn a threat to their predators, to which man belongs.

Population catastrophes in marine fish due to overnutrification may occur in coastal waters. On September 3, 1970, a fish kill in Escambia Bay, Pensacola, Florida, which involved the death of 10–15 million menhaden (*Brevoortia*) was one of forty-two kills that occurred from June 21 to September 4, 1970 (three in June, seventeen in July, sixteen in August, and six in September). The extent of these kills varies from hundreds, to thousands, to millions. Menhaden kills have been occurring every year during these months for at least the last ten years. All are due to overnutrification. Most of the kills occur in the small bays off of the big one. The algae and plankton populations grow rapidly due to overnutrification resulting from industrial discharge into the bay. When the menhaden come in to feed on the plankton, there is not enough oxygen for the large amount of fish.

The most extensive kill occurred in July, 1970, in one of the junctions of the Escambia River where it enters the bay. This was the only one in which almost all aquatic life in the area was killed (the others involved only menhaden). Although tests were run and the cause was undetermined, oxygen deficiency was thought to have been the important factor.

Menhaden are used for food, oil, livestock feed, fertilizer, and are a source of food for other commercial fish. The situation is not expected to improve until overnutrification is checked. The industries involved have been ordered to take corrective measures to halt pollution by January, 1973. The present conditions are the result of ten to fifteen years of misuse, so it is expected to take at least six years for any results to be observed.

Also, dumping of solid wastes into the open sea has in recent years taken place indiscriminately and in large quantities. At an FAO Conference in December, 1970, it was reported that practically all fish in the North Sea and eastern North Atlantic contain a particular type of chlorinated hydrocarbon, which apparently originates from industrial dumping and resulted in mass death of plankton and lowered condition of fish. The latter were found floating on the surface.

Industries, cities, and governments seem to believe that the absorption capacity of the sea is without limit. But the sea is not just a liquid. It is a life zone containing organisms essential for the well-being of the biosphere. Green algae, vitally important to the earth's oxygen supply, are sensitive to chemical poisoning. Continuously exposed to toxic chemicals, they get no chance to recover. At a still unknown degree of poisoning they will simply die in masses. We know that tiny amounts of DDT can disable the phytoplankton that perhaps produce 70 percent of the world's atmospheric oxygen. This suggests the possibility of an environmental change that could cause a sudden reduction of oxygen supply. Simultaneously terrestrial oxygen producers such as forests are disappearing. The actual elimination of marine algae and terrestrial plants on a scale never before known may create an ecological crisis.

The largest recorded disaster for sea birds occurred in 1969 off the northwestern coasts of Great Britain, where about 200,000 birds may have been affected. These birds—cormorants, shags, gannets, guillemots, razorbills, and puffins—were all fish-eating species. They were washed ashore dead and dying in September and October, 1969. In this case oil pollution was not the cause of the death of the majority of the birds. Some carcasses revealed fairly high levels of DDE (a DDT breakdown product) and of polychlorinated biphenyls, a plastic industry byproduct. However, the general emaciated condition of the dead and dying birds is lending weight to the theory that death was caused by food shortage or, in other words, by a catastrophic interruption of the natural cycle of events in the marine ecosystem within the area concerned. In such a situation of starvation the birds affected begin to absorb their accumulated fat reserves, where poisonous compounds are stored as a result of their previous feeding on fishes carrying toxic chemicals in their tissue.

The disaster of Great Britain is one of the many indications of how the marine ecosystem and food chains have been disrupted. This is again a lesson to man.

Another form of drastic environmental change is the continuous discharge of heated waters from atomic power stations. A relatively small increase of the mean water temperature in a given area will have tremendous consequences not only on some marine life species but on the entire ecosystem. The whole plankton productivity may be upset and many life cycles and food chains broken.

Radioactive waste dumped into the sea is another potential danger. It, too, may have serious consequences not only for marine life but also for man, who is the last link of many food chains. Radioactivity from nuclear fallout can be found in any fifty-gallon sample of water taken anywhere in the sea.

Despite the fact that the oceans cover almost three quarters of the earth, there is at present an almost total absence of international law. Except for a few fish conservation and fishery treaties and certain national laws, there is no efficient legislation or treaty governing the utilization of the oceans and the sea beds. Efforts have been made to ban nuclear weapons emplacements in the oceans, but the anarchy still prevails and the pollution hazards increase.

In April, 1971, the governments of Denmark, Finland, Iceland, Norway, and Sweden issued a call to end the dumping of industrial and chemical wastes in international waters. The other ten contracting states of the North East Atlantic Fisheries Convention were urged to ban discharge of wastes into the sea, particularly the North Sea. The five Nordic countries will also try to reach an international agreement to end pollution of the Baltic Sea.

This initiative is good, but it had already been taken by the United States, which has submitted a draft Convention on Ocean Dumping to the International Working Group on Marine Pollution established by the Preparatory Committee for the United Nations Conference on the Human Environment of 1972. Hopefully an international agreement will be reached at that conference.

There is an urgent need for more intensive research in marine habitats. The 1970's have been declared to be the International Decade of Ocean Exploration.

Marine Mammals

Man's destruction of marine vertebrates other than fish and birds has reached criminal proportions in the past, and the slaughter continues. Sirenians, sea otters, seals, whales, and sea turtles have for centuries been among the most mistreated groups of animals in the world. The fact that these animals, particularly all the larger whales and the sea turtles, are renewable natural resources of high economic value has not prevented man from reducing the populations of many species to the verge of extinction.

Marine mammals respond to harvest pressure and conservation measures in the same way as do marine fish. The experience with seals and sea otters parallels that of the Pacific halibut story. The largest seal rookeries in the world are situated on the Pribilof Islands of the Bering Sea, where the northern fur seal (*Callorhinus ursinus*) has its headquarters in summer. That is the breeding season, when the males gather large harems for mating and the females give birth. In winter these seals swim southward in the Pacific, feeding on fish, returning to their islands in early summer.

That the northern fur seal had a high commercial value was soon realized after the enormous herds comprising several million seals were discovered on the Pribilof Islands in 1786. An indiscriminate killing began, and within twenty years the herds had been reduced so heavily that the species was near extinction. In 1834 the Russians, who at that time owned the islands, regulated the hunting so that only the killing of mature bulls was permitted. After the United States bought Alaska from Russia in 1867, even stronger conservation measures were introduced in the Pribilofs, but this regulation was sabotaged by indiscriminate shooting in the open sea. The harvest on the Pribilof Islands dropped from the permitted 100,000 fur seals in 1870 to only 12,000 in 1910. The population decreased from between two and three million in 1864 to about 125,000 in 1911. In that year an international treaty banned all hunting of fur seals in open waters. This measure saved the northern fur seal. By 1943 the annual harvest exceeded 100,000 animals, though killing of only three-year-old males without harems was permitted. By 1947 the herd had regained its original size of about three million. In 1958 the summer population was estimated at about 1,800,000 on the Pribilof Islands alone. The annual crop from this population is worth about two million dollars.

The sea otter (*Enhydra lutris*), which has a very valuable pelt, offers another dramatic story. This species was discovered in 1741, ranged along the northern Pacific coasts in the New and Old Worlds. Russian ships began to hunt sea otters in incredible numbers. One ship alone could collect as many as 5,000 skins during a season. Not until the 1860's did the depletion seem to have become noticeable. When the United States acquired Alaska and the surrounding islands in 1867, the slaughter continued, but the number taken dropped every year. In 1910 protection was set up for the species, but the step appeared to have been taken too late. For many years no sea otters were observed. Then a small group was rediscovered. By 1935 this colony had increased to about 650 animals. In 1936 another group was found in the Aleutian Islands. Continuous protection helped the species to reestablish its population in fair numbers and to resettle formerly occupied coasts. At present, about 30,000 sea otters are alive and breeding, and the species can once more be cropped within limits, to be utilized as a valuable natural resource.

The examples of the northern fur seal and the sea otter clearly show how worthwhile it is to manage such a marine resource as aquatic mammals. Unfortunately man has done so only as the exception rather than the rule. His treatment of the large whales, by contrast, is a sickening recital, partly because of the cruelty of the catching methods, partly because it reveals an incredible human shortsightedness, egoism, and cynicism in the plundering of an exceptionally valuable natural resource.

The larger whales are one of the wonders of the globe. The blue whale is the largest animal that has ever existed on earth. The logging of the last giant sequoia would cause humanity to cry out in rage against such vandalism, but the fact that in recent years the blue whale has been brought to the verge of extinction by three whaling countries—Japan, Norway, and the Soviet Union—seems to concern only the conservationists.

It is strange that the other nations of the world are allowing these three countries to exploit ruthlessly a renewable natural resource that they do not own, for their own profit and without regard for the future. The whales are an international property. They inhabit the great seas and normally concentrate in the Antarctic and Arctic oceans. Almost all effective conservation measures that international organizations have suggested have been rejected by one or more of the three whaling countries, and the whaling goes on, depleting one species after the other. Apparently these countries deliberately plan to take as much as they can from the whale populations, until the business ceases to give a profit, heedless that several species will become exterminated. This cynical policy is present-day buccaneering—a plundering of a global resource belonging to every nation.

The cruel method of killing whales by projectiles exploding within their bodies is a subject beyond the scope of this book, but a few facts deserve mentioning. A wounded whale suffers hours or a whole day of terrible agonies before dying. Whales have the same nervous system as man and a degree of intelligence that may be second to man, or comparable to that of the apes. The pain felt by whales wounded by internal explosions is presumably proportional to the suffering men experience from the same type of wound. The modern human methods of killing whales are among the most shameful manifestations of civilized man's cruelty.

Several whale species dwarf all other living or extinct animals of the world, including the giant reptiles of the past. The blue whales measure more than one hundred feet in length and weigh more than 323,000 pounds. This species and all other baleen, or whalebone, whales are surprisingly disproportionate to the size of their food. They are specialized plankton-feeders with physiological adaptations that enable them to harvest plankton directly by filtering the almost miscroscopic food. Through evolution these whales have long since solved a problem with which man is still struggling. The two largest fish, the whale shark (*Rhineodon typus*) and the basking shark (*Cetorhinus maximus*) as well as the manta (*Manta birostris*), largest of rays, have comparable filtering systems. The method employed by the larger whales and other large forms of marine life to feed directly on tiny planktonic organisms is a highly efficient and economical way to build up the tissue of their huge bodies, because there is no energy loss by passing through other food chains. This helps to account for the high economic value of whales. Oil is the most important commercial product obtained from them. It is used in the manufacture of edible fats such as margarine, as well as of soap, ointments, cosmetics, and glycerin. Whale oil is also used to lubricate moving parts of chronometers and other delicate instruments. Many other products are derived from the whales, of which meat is not the least important. Whale meat is sold as food in North America, Europe, and Asia, particularly in Japan and China. Ambergris is perhaps the most famous of the valuable whale products. At present the annual crop of all harvested species of whales in the Antarctic provides more than half a million tons of meat and edible oils. This amount may seem impressive, but only thirty years ago more than that was obtained from one species alone in the Atlantic.

Before 1920 the Antarctic population of the blue whale (*Balaenoptera musculus*) was estimated at 200,000 animals. In 1930 it was down to 40,000, with an annual catch of 29,000 animals. This was 82 percent of the whole whaling catch in the Antarctic. By 1959 the blue whales comprised less than 5 percent of the catch. The season 1959–60 yielded 1,239 blue whales, 1963–64 only 112. By then the population estimates were down to 650, with a maximum of 1,950 animals. In 1964–65 only one blue whale was caught in the Antarctic. In 1966 the species was protected against hunting from ships in the Antarctic, but not from land-based stations and not in other seas, where at least 131 blue whales were killed that year. In 1968 shore stations also ceased to hunt blue whales.

When the blue whale became depleted, the whaling companies turned to the next species in size, the fin whale (*B. physalus*), which may exceed eighty feet in length and weigh up to 176,000 pounds. During the 1930's the Antarctic population of fin whales was estimated at about 200,000 animals. In the season 1956–57 it was down to less than 102,000 individuals, with a catch of 27,757. In 1963–64 the corresponding figures were 32,400 and 14,422 respectively. By 1966–67 less than 3,000 fin whales were taken. The whaling companies then concentrated on a third species, the sei whale (*B. borealis*), and now its numbers too are dwindling alarmingly.

These three species of whales are by no means the only ones depleted by man. It took whalers only about a hundred years to exterminate the whale populations that existed in Arctic waters in the seventeenth century. The Greenland right whale (*Balaena mysticetus*) was hunted there almost to extinction. The slaughter went from one ocean to another until the whaling industry moved to the Antarctic, where it is now carrying on its tradition of ruthless exploitation. Recently whaling has been reintensified in the North Pacific due to the decline in the Antarctic, and a new overexploitation is proceeding there.

The humpback whale (*Megaptera novaeangliae*) was not hunted very much before the twentieth century, but in a few decades of whaling this species was almost wiped out. Its numbers fell so rapidly that even the whaling countries recognized that protective measures were necessary. Protection was established in 1938, and in 1949 hunting started again. The catches soon dropped drastically, and by 1965 the humpback was given protection in all seas from hunting by ships, but land whaling stations continued to catch humpbacks. In 1965, for example, 118 humpback whales were caught off Peru alone. However, in 1968 land-based whaling stations ceased to kill humpbacks.

Species by species, region by region, the whale slaughters and depletions have been repeated. It is not feasible to continue the tragic story by giving an account of the other species of whales involved in this sordid chapter of man's relations with marine resources.

The whale recital emphasizes the need to take prompt action for a rational management of the world's whale populations by an effective international body other than the International Whaling Commission. The latter organization has clearly demonstrated a failure of responsibility and a lack of capacity to deal with one of the most important natural resources of the seas. The whaling companies seem to dominate the commission, which even disregards unanimous recommendations of its own scientific committee. It has ignored all pleas for another policy. The only way to save the larger whales from extermination, and the whaling industry from total collapse, is a drastic reduction in the catches, including total protection of several species. This is recognized by all experts, but the three nations actively engaged in Antarctic whaling continue their catches, apparently blind to all arguments except their own immediate profit.

Obviously, the main reason for this continued plundering is that nobody is territorially and juridically responsible for areas in the open sea, which leads to a situation where nobody cares about a sound management and wise utilization of a renewable resource. Instead, by jealousy and greediness, the whaling nations try to get the largest share of the capital in the shortest possible time without planning for the future. Incredible as it may seem that such depredations can occur in the second half of the twentieth century, the facts are there.

The positive examples of effective conservation measures for fur seals and sea otters that have been cited were possible because these resources were restricted geographically and belonged to a few responsible countries that were able to reach agreement on managing their marine resources. It should be possible to accomplish similar results in whale conservation at sea if the nations involved could trust each other and show a minimum of foresight. Biologically all the species of whales could be restored to their former abundance. The basic conditions for such a comeback exist, provided that the population level of the blue whale is not below the point where a normal reproduction rate can be maintained. It would be economically, scientifically, and aesthetically inexcusable for this generation deliberately to exterminate the whales. If this should happen, future generations of human beings would never be able to understand why. They would have to ascribe such stupidity to some unaccountable mental defect in the human beings of the twentieth century.

Marine Reptiles

Differing from most marine mammals, the sea turtles belong chiefly to tropical waters. Like the whales, most of the six (perhaps seven) species are worldwide. Despite intensive research during the last decades, we know relatively little about the biology of the sea turtles. They spend most of their lives in the open sea, though some species are confined to shallow coastal waters. However, all species have to reproduce on land, and therefore all of them occur, at least temporarily, in coastal waters.

The necessity for marine turtles to go ashore during the breeding season to lay eggs makes the adult females very vulnerable, and that is also the case with the eggs and the newly hatched young. In addition, exploitation of shore areas and alterations in shorelines that serve as breeding sites for sea turtles reduce the possibilities of reproduction. For centuries a careless exploitation of sea turtles has been taking place, with the result that several species have been severely depleted in many areas where formerly they had been abundant. About 1.6 million green turtle (*Chelonia mydas*) eggs are collected in the Delta of Irrawaddy in Burma each year; 1 to 2 million are gathered in Sarawak, and 1 million in the Philippines.

Sea turtles are exploited mostly for their importance as food. Here again is a highly valuable group of animals living in the sea, where without competing with man they convert plants and small animals into excellent meat, oils, and tortoiseshell. The green turtle is herbivorous, grazing on manatee grass and turtle grass. It furnishes meat of unsurpassed quality and could, if managed properly, constitute a very important food resource in tropical and subtropical countries. Its protein supply potential has been underestimated, and in most countries where it is available, almost entirely neglected by the governments, despite the fact that for centuries the sea turtle has been a considerable part of the diet of human coastal populations and probably played an important rôle in the settlement of the Caribbean. The former abundance of the green turtle in the Caribbean must have been enormous. Turtling was an industry not only there but also in the whole area of the Gulf of Mexico and the Atlantic north of the West Indies to Bermuda.

All governments concerned need to realize that they have a responsibility to conserve and manage the world's most valuable reptile, which the green turtle certainly is, so that it can be used rationally as a food resource on a sustained yield basis. The first step would be to restore the populations to the optimal level that the environment can produce. Artificial hatching and raising are possible and have already been undertaken successfully, and regular releases of young green turtles into the sea are now made in many parts of the world. But this work is only a start.

The hawksbill turtle (*Eretmochelys imbricata*) occurs in shallow coastal waters of the tropics and subtropics. It feeds on algae, seaweeds, and various invertebrates. It produces good meat and valuable shell and has great economic importance. The loggerhead turtle (*Caretta caretta*) is also a valuable species, though its flesh is not as well flavored as the hawksbill's. Its eggs, highly esteemed, are collected annually in high numbers, and its oil and carapace are of value. This species feeds chiefly on crabs and conch, also on other invertebrates and at least occasionally on grass.

The ridley turtle (*Lepidochelys kempii*) likewise has great economic value, but of all the sea turtles none is more in danger than this species. Like all the sea turtles, it was once very common. Veritable armadas swarmed ashore to lay their eggs on Mexican beaches. Severe depredation sharply reduced the number of the ridley to a fraction of its former abundance. At present only one breeding site is known, where about three thousand ridleys congregate for egg-laying. The warana turtle (*L. olivacea*), distributed in the Pacific, Indian, and Atlantic oceans, is also a useful species. The majority of the waranas come ashore to breed in Surinam (Netherlands Guiana), where practically all the eggs laid are taken by man. The leatherback turtle (*Dermòchelys coriacea*) is a giant that reaches 8 feet in length and weighs more than 1300 pounds. It dwells more in the open sea than any other species of sea turtle, living in the tropical and subtropical areas of all oceans. It apparently builds up its huge body by an omnivorous diet of plants and invertebrate animals. Its eggs are prized food items, its meat is edible, and its oil and shell are other products of value. The collecting of eggs for commercial purpose has proved disastrous for the leatherback turtle, and it is badly in need of protection. Finally, there is probably also a seventh species, the flatback turtle (*Chelonia depressa*), which occurs in Australian waters. Its taxonomic status is a matter of controversy. Some scientists regard this form as a race of the green turtle.

This brief survey of the six or seven sea turtles shows that they are of economic use to man in many ways, chiefly as a potential food resource of high protein value. In addition, some species are esteemed as a delicacy and can be developed, as once they were and in some areas still are, into an important export article. It would be worthwhile to pay more attention to the sea turtles in all development plans in tropical and subtropical countries facing the sea. An international management plan should be worked out, region by region. But to arrive at such a plan requires intensified research and, in the meantime, vigorous protection on the beaches where sea turtles reproduce.

An international program of action for the development and utilization of sea turtles as a resource can be expected to substantially benefit the economies of the nations involved. Without such a program, including immediate conservation measures, sea turtles will have no stable future as a marine resource and may rapidly disappear through ignorance, greed, and vandalism. A sea turtle conservation program would be in reality a development program for an important potential economic resource. Sources of aid to developing countries may well give serious consideration to providing financial assistance for such a program. Even local programs would have a potential tropics-wide application and should be given high priority.

Estuaries and Brackish Waters

Tidal estuaries and brackish water habitats produce special environmental conditions that vary greatly from one area to another, depending on tides and freshwater outflows, temperature and salinity, bottoms and vegetation, and so on. In the tropics mangrove growths represent a very special environment, particularly below the water surface where the tangle of mangrove roots creates a peculiar underwater world with an extraordinary gathering of marine and freshwater animals. Usually estuaries and mangrove forests are regarded as unproductive wastelands, suitable for dumping industrial and urban wastes. The contrary is true. Estuaries are often among the world's most fertile areas.

Coastal shallow waters and estuarine wetlands are often subjected to physical alterations besides being heavily polluted. Draining, dredging, filling, and waste dumping of such coastal waters and lands are not seldom highly uneconomic and can greatly reduce or entirely eliminate the food base for a high number or even all organisms in such wetlands, with negative repercussions also to human economic interests.

Interesting basic research on the role of the tidal marshes in estuarine productivity has been carried out by Professor Eugene P. Odum of the University of Georgia. The area in that state under study consists of as yet unpolluted streams, marshes, flats, creeks, and bays of an estuarine ecosystem. A basic factor underlying the productivity of the entire system is the tidal action in nutrient cycling. The natural productivity is of such a magnitude that management emphasis has to be on utilization rather than production. According to Odum, the fertility and productivity of estuaries—measured by ability to fix organic matter from sunlight—rank with coral reefs and intensive rice and sugar cane cultures. In terms of dry weight fixed annually per acre, Odum's estuarine figures from Georgia exceed ten times the corresponding values for ecosystems such as grasslands, forests, eutrophic lakes, and ordinary agriculture. The data he obtained in Georgia are believed to be applicable throughout the Atlantic and Gulf coasts of temperate and subtropical North America. If this is true, estuaries in other areas of the world should likewise be highly productive.

Odum estimates that the Sapelo estuarine marshes in Georgia have a gross primary net production of about 2,000 grams of dry matter per square meter per year. This is 10 tons per acre. The highest yields of wheat and corn obtained in northern Europe are about 7 tons of dry matter per acre. World average net production of wheat is about 1.5 tons of dry matter per acre per year. In terms of potential food energy usable by organisms, the net production of the Georgian estuaries investigated by Odum is apparently in the order of 8,000 calories per square meter, or about 32 million calories per acre per year. To judge what this figure means in relation to the food requirements of man: on a world population basis, the mean quantity of food to keep a human being healthy is estimated to be 2,750 calories a day.

Several major mechanisms explain why estuaries are so fertile and productive. Odum points out that the mixing of waters of different salinities produces a vertical mixing of nutrients. Instead of being swept out, the nutrients move up and down, cycling rapidly among organisms, water, and bottom sediments. Secondly, the back and forth tidal flow continually supplies nutrients and oxygen, and automatically removes waste products, so that organisms need not use their own energy for these purposes. Odum's data reveal the remarkable efficiency with which a tidal estuarine ecosystem functions. They indicate how careful man should be in altering and managing estuaries so that their finely articulated year-round crop mechanism will not be destroyed. Perhaps the most important of Odum's findings on estuaries, next to how extraordinarily productive they are, is that the entire estuary marsh complex—sounds, creeks, mud and sand flats—must be considered as a unit, regardless of whether it is one or one hundred miles wide.

The prime importance to humans of the tremendous productivity of tidal estuaries seems to be as a source of seafood. Estuarine wetlands continuously fertilize the pastures and bottoms of the sea and are in this way a basic link in the marine food chain. It is certainly more useful for man to leave potentially useful tidal marshes and estuaries intact, instead of converting them into less productive lands for agriculture. Man needs more high-quality, self-produced, protein-rich seafood rather than artificial lettuce crops grown at heavy taxpayer expense. Here again it is better to cooperate with nature than to combat it.

Apparently estuaries and brackish waterways well protected from the open sea are a natural resource that is entirely neglected or badly used in most places where they exist. Brackish inshore water may even be farmed. The potential value of such agriculture may be judged by comparing the productivity of already existing commercial brackish ponds and conventional farmlands. In tropical Asia such ponds yield from 160 to 1,500 pounds of fish and prawns per acre per year. The annual total edible protein, expressed in dry weight, is up to 81 pounds per acre in brackish ponds, while the grain produced per acre under average agricultural conditions in the United States will feed enough livestock to yield about 21 pounds of protein food a year.

One must bear in mind that modern agriculture is based upon intensive research into all its aspects, while aquaculture has virtually no formal study underlying its methods. Yet there are ponds in southeastern Asia where, in addition to the rice crops, 1,500 pounds of prawns, net weight, are harvested annually. In such cases brackish-water farming yields at a much higher rate than agriculture and even a better fishing rate than that obtained by commercial fishermen. Undoubtedly the protein harvest from brackish waters can be multiplied by management measures that are much less costly than what is required to maintain the productivity of conventional farmlands.

But quite aside from the possibilities of a managed agriculture, tidal marshes bordering rivers, estuaries, and bays, with their tremendous productivity, are among the most important and valuable wetlands by virtue of their function in maintaining productive fisheries along the coasts. The economic role such ecosystems play is primarily as fertilizers of the "pastures of the sea." They also produce food "directly" for a wide range of brackish-water organisms: plants, invertebrates, fish, ducks, geese, and so on. Therefore it is of utmost importance for man to preserve tidal marshes and manage them wisely. The revenue will be immense and the benefits far-reaching, particularly for coastal fisheries.

Coral Reefs

In a biological sense coral reefs are a world asset. They are important economically because the productivity of their organisms is high and contributes to several food chains in the ocean. Their fishery value is high, both directly and indirectly, for they provide food, shelter, and spawning sites for marine life. In addition, they have mineral and recreational values.

As structures, coral reefs are vulnerable to human destruction. For example, the Great Barrier Reef off eastern Australia, one of the wonders of the world, is subject to several hazards. It consists of 4,000 to 5,000 individual coral reefs that extend about 1,200 miles along the coast from 10 to 200 miles offshore. Prospecting for offshore petroleum deposits may lead to oil drilling activity in those waters, with consequent risks of spillage that would be highly dangerous to the marine life in general and to the coral reef habitats in particular. One major oil spill has already taken place at the Great Barrier Reef. Mining activities for coral detritus may also have fatal effects. Finally, many reefs are beyond territorial limits and therefore not protected by national laws. Needed conservation measures involve complex international legislation and conventions. Before such laws and agreements exist, coral reefs may be ruthlessly exploited.

The trade in corals and shells as curios for visiting tourists has grown to such dimensions that it is now a threat to the living species and entire coral reefs. Some beaches, littorals, and reefs in the tropics have been virtually cleaned by collectors for commercial purposes and are now like lifeless submarine deserts.

Here again, conservation efforts and management planning of coral reefs, based on ecological research, would in the long run benefit mankind in the form of dividends from a chain of marine renewable resources.

Recently another deadly menace to coral reefs has taken great proportions. The crown-of-thorns starfish (*Acanthaster planci*), which since millennia has been a normal inhabitant in coral reef habitats, where it feeds on living coral, is having a population explosion in numerous areas. This phenomenon is probably caused by man-induced changes in the marine environment and balances.

Destruction of corals by this starfish was first reported on a major scale in the Great Barrier Reef in 1963. Later the same damage was observed on reefs off some West Pacific islands, where the most serious infestations were found on Guam, the Marianas Islands, and the Caroline Islands. In 1968–70 more and more Pacific coral reefs became infested by *A. planci,* and its killings of corals is locally catastrophic for the reefs as a living community. At present the crown-of-thorns starfish is found in large numbers not only in the areas mentioned above but also in Palau, New Caledonia, New Britain, Fiji, Samoa, the Tuamotus, the Hawaiian Islands, as well as Borneo and Thailand. This means that the plague has also affected areas in the Indian Ocean.

In 1968 only a few crown-of-thorns starfish were observed in the Red Sea, but in 1970 there were thousands of them. However, the only "plague" concentration in the Red Sea was in the most polluted reef area near Port Sudan.

The damage done by *A. planci* is not only threatening the corals but it also upsets seriously the marine life system of reefs around many islands and atolls. Invertebrates and fish that have adapted to life in a living coral reef cannot adjust to a radically changed environment. This elimination of species affects the whole food chain. In dead reefs, recently killed by *A. planci,* larger food and game fish are almost totally absent and even deep-sea fish populations may be affected by this breakdown in the food and life chains.

If current trends of coral reef destruction continue unchecked, results could include fish shortages for human island populations, who rely on fish for protein, loss of revenue from tourism, and possibly erosion on some low-lying islands now protected by coral reefs. Moreover, pelagic fishes which use coral reefs for spawning and the first life stages will certainly also be affected.

The cause of the tremendous increase of *A. planci* is still unknown. Obviously some environmental change must have taken place which has favored this starfish and broken the balance of the marine ecosystem. Several suggestions have been made: dredging and blasting of reef areas have decimated creatures normally preying on the larvae of *A. planci;* shell collectors have substantially reduced the number of the giant triton snails (*Charonia tritonis*), which prey on adult starfish such as *A. planci;* and concentrations of organo-chlorines and other man-made pollutants have been increasing in the marine environment during the past few decades. These pollutants might have eliminated planktonic predators, decreased the zooplanktonic standing population, or reduced the reproductive capacity of the zooplankters so that the zooplanktonic community can no longer expand rapidly enough to check the influx of large swarms of *A. planci* larvae during spawning seasons. . . . Another hypothesis suggests that predators of the adult *A. planci* (i.e., tritons) might accumulate organochlorines in their tissues and that this might reduce the reproductive capacity of the predators. Subsequent release from predation might lead to the current population explosions.

The finding of concentrations of crown-of-thorns starfish in the Red Sea where ships queue to await permission to enter Port Sudan harbour and where condemned foodstuffs are dumped and tanks cleared could be of some significance.

Mangroves

Mangroves occur along marine shores of tropical and subtropical seas but also grow in deltas, estuaries, brackish water swamps and up rivers as far as the tide carries salt water. Locally mangroves are even found beyond the intertidal zone. They are, in fact, land plants which have evolved a high salinity.

Mangroves are important land and island builders which not only extend coasts but also protect them from erosion and hurricanes. Moreover, as said in a preceding subchapter, mangroves provide habitats for numerous marine organisms and constitute feeding, breeding and nursing grounds for many species of fish and crustaceans. Also birds and reptiles utilize mangrove habitats.

During the last decades several tidal-waves and hurricanes have caused disasters along tropical coasts where man has very widely removed the mangrove vegetation in order to convert the areas to paddy fields for rice production. This is immediately followed by the establishment of fast-expanding villages. This was, for example, the case in Bangladesh in November 1970 where from 300,000 to 500,000 human lives were lost. The fact that the hardest hit regions in the Ganges Delta were areas which had been subjected to extensive clearing of mangroves may indicate that man was at least partly responsible for the effects of the 1970 hurricane disaster. Therefore, the presently accelerating conversion from mangroves to open lands which become unprotected from ocean surges generated by hurricanes, seems to be ecologically unwise.

CHAPTER FOUR

Fresh Water

Without water, plants and animals cannot survive. All living organisms contain water and use it externally. Water is our most precious mineral. From the beginning of history it has been a key to civilization and development. It is the largest single controlling factor in the growth of populations. The available water supply is a boundary line beyond which no society or nation, agricultural or industrial, can go. Perhaps no greater conservation problem faces mankind today than that of keeping the waters clean and maintaining adequate and qualitatively useful supplies of this natural resource.

Although fresh water is self-perpetuating and universally distributed, its regenerative power is complex and in many ways interrelated to the earth's physical and biotic elements. This means that man's exploitation of renewable natural resources may also affect the rainfall. In fact, drought is spreading. Regions that once enjoyed abundant annual precipitation are at present facing water shortages. This is happening all over the world, but not where large forests still exist. A climatic change may be involved, but by itself could hardly have achieved catastrophic effects in so short a time. Man has accentuated the negative consequences by destroying renewable resources, making the environment less resistant to damages from drought.

In many parts of the world, periods of drought are normal phenomena, which can prevail for decades or even centuries. This is the case in Australia, for example. But life goes on there, having adapted to the long-term climatic fluctuations. In other regions where man has drastically changed the natural scene, removing the protection against drought, organisms, including man, have difficulty adjusting to the new conditions.

Local or regional drainage, irrigation, deforestation, agriculture, or pasturing may not have any long-term value at all if, as often happens, the hydrography of the ecosystem is seriously affected. Exploitation measures undertaken at the earth's surface may have drastic effects on the position and seasonal range of water table fluctuations, on the direction and quantity of the water movement, on evaporation, the runoff of rainfall, the effects of flashflooding, and so on. Ultimately the rainfall itself may be adversely affected, and the local climate actually changed.

When discussing water as a resource, it is necessary to visualize that it is one element in a complex watershed system also involving air, land, vegetation, and animals. The water circulation from clouds to lands and back to the atmosphere is another complex system of transpiration, evaporation, and precipitation, partly determined by the character of the drainage basin. Its interrelationships dramatize the environmental complexity and the impossibility of considering water as a separate, independent element.

As a mobile resource, water rapidly affects the ecology of large regions. If it is polluted or contaminated, the damage is quickly spread over wide areas. Therefore, the management of freshwater ecosystems is of fundamental importance for the maintenance of aquatic quality, on which so many other resources depend.

Ground Water Table

An American scientist has estimated that in the United States the ground water occurring at depths less than 2,500 feet is equivalent to the total of all refilling by natural flows during the last 160 years. If this is true, the water comprises a sizable reserve. But in many areas of the world where the soil has been overutilized, the vegetation cover has been eliminated, or marshes and lakes have been drained, the subsoil water level is rapidly sinking or even disappearing. Man is destroying the ground water resources that in many regions could be his ultimate reserve in periods of extreme droughts. Wherever human populations are rising, and particularly in industrial countries, man's water needs are inexorably increasing. But the quantity of water that nature produces will remain almost constant, or may be regionally reduced due to bad land use by man.

Present use of water exceeds or will exceed the supply in many areas of the United States, for example in much of California, the arid Southwest, and parts of the Middle Atlantic States. Also ground water resources are tapped, which indicates how desperate the situation is.

In temperate, once glaciated regions, eskers—ridges of gravelly and sandy drift formed by streams flowing under glacial ice—are particularly valuable. They usually hold great quantities of subsoil water and therefore regulate the ground water table of a whole region. Hence exploitation of eskers to obtain gravel or other deposits may have disastrous effects on the hydrography and economy of a large area within a drainage basin.

The sinking of the ground water table in coastal regions may result in an inflow of salt water from the sea, with serious environmental consequences for man, animals, and plants.

Water is not everywhere a renewable resource. In many present desert areas, particularly in Africa, it is possible that the underground sources of water are derived from a wet climatic period of 7,000–12,000 years ago and, therefore, are not inexhaustible.

Water Supply for Living Beings

Plants and animals inclusive of man need definite quantities of sufficiently clean water in the right place and at the right time of the year to be able to live and reproduce. In addition to these necessities are man's domestic and industrial water needs, which are tremendous. Millions of people living in the large cities of the world count on water coming automatically from the tap at any instant they turn on the faucet. Few of them are fully aware of how vital the water is to human life: that it requires 50 gallons of water to produce one cob of corn, 37 gallons for one slice of bread, 1,400 gallons for a helping of potatoes, 1,600 gallons for one pound of steak. To produce a ton of wheat, the soil requires about 44,000 gallons of water. To produce a single dollar's worth of steel takes 60 gallons; of paper, 300 gallons; to produce a gallon of oil from coal takes 600 gallons.

The daily available water supply flowing in the rivers of the United States has been estimated at between 1,100 and 1,300 billion gallons a day. Of this quantity between 560 and 700 billion gallons a day are expected to be captured in dams, reservoirs, and other water reserves. Currently (1971) about 355 billion gallons a day are used in the United States. By 1980 between 570 and 600 billion a day will probably be needed and by the year 2000 the daily consumption may reach 900 to 1,000 billion gallons—between 80 and 90 percent of the daily river flows. It is uncertain if so much water can be taken from the natural flows. If so, the same water will have to be used many times for various human purposes, and each time be returned as clean water. Another procedure is to adjust the size of human populations and needs to the water volume capacity of various catchment areas—a measure that will have to be taken sooner or later.

Lakes, Marshes, Bogs, Rivers, and Deltas

Wetlands of various kinds represent great potential assets. They have long since been so regarded, but for the wrong reasons. They were considered to be economic reserves that could be drained to agricultural lands when population pressure made it desirable to increase the arable acreage. It is true that some drainage schemes have been successful, but the general assumption that any wetland—a shallow lake, a swamp, a marsh, or a bog—is suitable for agriculture has led to destruction of useful aquatic resources and brought about economic fiascos on a gigantic scale. In reality, wetlands are assets precisely because they are wetlands. If they are kept as such, both direct and indirect benefits will accrue.

The claim is often made, with the Netherlands as example, that immense benefits are gained from drainages. But the remarkable reclamation projects in the Netherlands have been a successful conversion from sea to land—quite a different operation from drainage of a freshwater habitat. Marshes and other wetlands were secondarily involved in the Dutch scheme, but many of them have been converted into polders, themselves a kind of wetland that is managed to yield a variety of natural and cultivated plant and animal crops.

Left in their natural state, wetlands are a highly productive natural resource of importance, not only locally but often for the whole surrounding region and even beyond. They are used for breeding, feeding, resting, moulting, and wintering by migratory birds—geese, ducks, coots, swans, and waders—that as food or recreational resources may be of importance to people living on more than one continent. Wetlands production of fish may benefit a large proportion of the human population. They also supply other kinds of food at a high rate of productivity. They play an important recreational role and serve as reservoirs. Their greatest importance may be as a hydrographically stabilizing element within a drainage basin, watershed, or ecosystem. Wetlands function as sponges, accumulating excess water during floods, storing it for months, and distributing it during drier periods. They are important for the maintenance of the ground water table in vast subsoil areas. They are essential for the welfare and fertility of a country, preventing catastrophes resulting from violent floods and inundations and from the drying out of lands.

From the human point of view wetlands are not always beneficial, because in the tropics they are often centers for breeding mosquitoes that carry malaria and other diseases to which man is susceptible, and for harboring bilharzia, a fatal human disease.

Lakes represent a wide spectrum of aquatic habitats within each climatic region. Many lakes, both in tropical and temperate areas, are extremely productive and among the most important food resources of man. The food chains in lakes are as complex as in the sea, in some cases even more so. In Lake Edward, situated on the equator between the Congo and Uganda, a direct link exists between terrestrial grasses growing on the savannas surrounding the lake and the aquatic organisms in the lake. These plains are grazed at night by thousands of hippopotamuses (*Hippopotamus amphibius*), which occur here in a population that is the densest in Africa. Each hippo consumes 90 to 130 pounds of grass during twenty-four hours. The animals spend the days in the water of the lakes and the surrounding rivers, where practically all their excreta are left. In this way the hippos fertilize filamentous algae, which are remarkably abundant in Lake Edward. These growths and their attendant zooplankton are eaten by several species of the fish called breams (*Tilapia*). These fish are more numerous in Lake Edward than in any other lake and are the basis for an extraordinarily continued fishing yield. They also serve as food for several predatory fish, which add much to the Lake Edward fishery harvest. As *Tilapia* species is qualitatively the finest food fish of Africa and rich in proteins, the revenue of the Lake Edward fisheries is spectacular. The catches are distributed over vast regions in the Congo and Uganda.

The Lake Edward example illustrates the complexity of an ecosystem comprising savannas, rivers, and a lake, as well as a food chain, or more precisely, a flow of energy, consisting of terrestrial vegetation–a mammal–excreta–bacteria–aquatic vegetation–zooplankton–fish–fish–birds and mammals, including man. The existence of a dense hippo population is a condition for the enormous productivity in Lake Edward and the human fishery there. The principal reason why the densest hippo population of Africa has developed undisturbed on the shores of Lake Edward is that the area is located in Congo's Albert National Park, Africa's oldest national park (created in 1925), where the hippo and other animals have been totally protected since the 1930's—again showing how profitable it is for man to cooperate with nature and let an ecosystem function normally.

The value of marshes and swamps as safety valves in the hydrographic system of a watershed area and drainage basin cannot be overestimated. In addition to their ecological significance, marshland areas have direct economic values: for reed culture, which is of extreme importance in many countries, haymaking, moss harvest, fishery, hunting, and open air recreation. For many areas in different parts of the world, the recreational activities on and about lakes and marshes are the chief source of income. For the most part, recreation need not clash with conservation and management of aquatic resources. On the contrary, the recreational utilization of a wetland is often in direct proportion to its importance as a wildfowl habitat. In the United States alone, resource-oriented recreational visits to bird marshes and such other wildlife areas as national parks, national forests, and national refuges totalled 183 million people in 1960–61. Recent economic studies of the Horicon Marsh in Wisconsin illustrate the trend. In 1961, some 44,500 goose hunting trips were made into the area, while 75,800 people came just to watch the geese from the highway, which crosses a section of this marsh. In 1963, the record shows 67,367 made goose hunting trips to Horicon Marsh, and 202,500 visitors came to see the geese.

This trend is confirmed by figures for North America showing an increase of people deriving enjoyment from a nonconsumptive use of wildlife. Since the early 1960's there has been a decline in the number of those who utilize the wildfowl resources by shooting for their own satisfaction, as is the case with hunters. While this is the trend, it should be said that there is no harm in hunting, if carried out correctly on the right species in the right season in the right area. Hunting is a legal form of recreation and of utilizing a natural resource. However, nature-minded vacationers contribute more than hunters to the economy of a marsh area and the surrounding region. Bird watchers outnumber hunters, stay longer in the area, spread their visits over the year, and do not kill that which is the area's attraction. Thousands of people can watch the same bird, while only one hunter can kill it.

Bogs too have been considered wastelands. They usually develop in cold temperate areas, but exist also on tropical mountains. Compared with marshes and ponds the productivity is low, but as a part of a temperate ecosystem bogs may play an important role and be of economic value.

Streams and rivers vary greatly in volume, velocity, turbidity, transparency, temperature, chemical composition and other factors, but they are all running—an important habitat character. Running waters are ecologically and economically important for many reasons. The direct action of a current often determines what can live and be produced in a river. Streams form ways for dispersal and migration of life forms not only within the river system and its lakes but also between fresh water and the seas. Such water routes are essential for the existence of fisheries of high economic value in both the sea and fresh water. Such migratory fish as various species of salmons, trouts, sturgeons, and eels, which shift from the rivers to the ocean or vice versa during various life stages, would be wiped out if they were prevented from travelling freely in running waters. Many of these fish have a strong tendency to return, when sexually mature, to the brook in which they were hatched or spent their first period of life. For the eels it is the reverse: they go to the sea to spawn.

Many rivers transport enormous quantities of sediments such as pebbles and finer material. In Swedish Lapland a medium-sized river can transport between 5,000 and 10,000 tons of sediment during a summer's day, as against only a few tons in winter, altogether about 180,000 tons a year. Thus enormous amounts of material are transported downward to build up delta land. Eventually, luxurious vegetation begins to grow around innumerable channels, river arms, and lagoons. Many animals thrive in deltas, particularly wildfowl and fish among the higher organisms. Many deltas in various parts of the world belong to the most productive habitats from an economic standpoint. However, a delta ecosystem is dependent on the water volume running through it. If the volume is altered by human actions—for example, by control of water for irrigation or hydroelectric power hundreds of miles upstream from the delta—it may seriously upset the hydrography and productivity of a living delta.

Floods

Deltas, valley flats, and plains are often flooded seasonally, which usually is a condition for the maintenance of their fertility and sustained productivity, including that of cultivated lands. The natural flooding of the Nile Valley and delta affords an outstanding example of an area that for six thousand years has fed a relatively dense human population, despite its being surrounded by deserts. Periodically, every year through the centuries, the Nile has flooded its bed and fertilized the surrounding meadows and fields with fine layers of silt and organic material. In consequence the Nile Valley has remained productive for millennia, despite a very strong cultivation pressure. This is in contrast to other Mediterranean areas that today are struck by drought and erosion due to misuse of water and land.

The fantastically rich natural resource of the Nile Delta is presently threatened by the hydroelectric and irrigation scheme of the Aswan Dam, which will transform the natural delta into an entirely artificial system of canals. Although this is supposed to prove a long-term benefit to Egypt and its population, in reality it will prevent the fertilization of the lands by silt and other nutrient minerals. The silt load will be deposited behind the dam walls instead of in the delta. The immense sediment transports of the Nile will gradually fill up the dam. Moreover, a great amount of the water masses stored in the desert dam will evaporate without returning in the form of rains over Egypt. The agriculture of the Nile Delta, which during six thousand years has been the economic backbone of Egypt, will be changed into something that for the long run is highly uncertain. The Egyptian Aswan Dam project is not in cooperation with nature and will probably result in an agricultural failure, though the first years will show increased crops. But there is even more damage. When nutrients from the Nile were no longer washed out in the poor waters of the Mediterranean Sea, the food chain was broken—which led to a tremendous decline of fish populations.

The sardine industry in Egypt at one time amounted to 18,000 tons annually, worth about $7 million—almost half of the total marine catch. It now amounts to about 500 tons annually. This has caused a crisis for fishermen and reduced an important protein resource.

To kill a river system is a dangerous venture, particularly in a country which, like Egypt, is almost 97 percent desert.

The lower Nile Valley will lose annually about 100 million cubic centimeters of silt and sediment, which will accumulate behind the dam walls. The sediment load, previously carried by the Nile, will no longer consolidate the river bed and continue to build the delta. Instead, the flow of water free from sediment will, despite its low volume, increase its erosion power in the bottom of the old enormous river bed, where it will dig a deep and narrow canyon-like ditch. The water flowing there will be beyond the reach of the canals of the old irrigation system of the lower Nile. The delta will no longer be subjected to coastal erosion by the sea. A part of Egypt's commercial fishery was based on five brackish lakes and many shallow ponds, separated from the Mediterranean by bars of sand. These bars are now disappearing and the sea will flood the delta with salt water. The fertile land of the delta will be destroyed or simply vanish into the Mediterranean.

A further consequence is already evident: an increase of bilharzia and also of malaria, because the vectors of these diseases thrive in stagnant water. About 70 percent of the population of lower Egypt is now affected by bilharzia.

Indeed, the Aswan Dam project displays spectacularly how dangerous it is to neglect ecological warnings when interfering technologically with ecosystems. Actually the benefits of the $1.3 billion Nasser Dam seem to be cancelled out already by the negative aspects of this unnecessary project.

Floods, as everybody knows, are not always beneficial. Every year disastrous floods occur around the world. China in particular has long been the scene of tremendous inundations. In the United States the problem of managing the waters of the Mississippi and its tributaries is of national concern. Floods even seem to occur with greater frequency and strike with increasing violence. Catastrophes of such magnitude as the floods of Italy, which in 1966 severely damaged parts of Florence and other cities, and the one in the lower Danube area that hit Romania in 1970, are not known to have occurred previously in historic time. This is likewise true in Asia and North America.

Destruction of vegetation cover, combined with reckless ditching and draining, undoubtedly has led to increased frequency and severity of flood catastrophes. The hill and mountain forests of Italy and those surrounding the Danube basin are to a great extent gone. China, once largely covered by forests, is now only 9 percent forested. The tremendous landscape transformation in North America, particularly in the Mississippi Valley and basin, where thousands of lakes, ponds, and marshes have been drained, is the chief reason for the destructive floods of the present time. Floods indicate a misuse of water, which can hardly be changed for the better without a restoration of wetlands and forests. While such conservation programs are on the way in many parts of the world, they are costly, and up to now only a fraction of destroyed aquatic resources are involved in these restoration projects. Other forms of flood control measures are a more artificial kind, for instance, the construction of levees. They are hardly as effective as wetlands restorations and will often create new problems without solving the old ones. Flood control is another human activity that has to be carried out in cooperation with nature, not against it.

The usual way to destroy nature's own water control is to eliminate wetlands and forests, and then to control the consequent floods by building enormous dams at great cost. The dams in turn negatively affect all the surrounding habitats and may prevent the water from reaching dry areas that need it badly. This then necessitates artificial carriage of water to dry regions at the expense of other natural resources. A vicious circle is established, consisting of man-made actions that in the long run have adverse effects. Water resources have to be managed and restored with the participation of ecologists and hydrologists, not by engineers and agriculturists alone.

Productivity

All persons involved in land use planning will do their work better if they understand that wetlands are a renewable natural resource comparable to forests, grasslands, and farmlands. Wetlands' productivity, their versatile utilization, and their hydrographic significance justify every effort to conserve and manage this resource properly.

The productivity of wetlands is probably best illustrated by fish. In central Europe, for example, it has been found that fish may provide a yield of human food per acre substantially greater than that obtained from wheat. Fish farming is a potential in many waters not presently used for that purpose. Some of the less esteemed fish like ide (*Leuciscus idus*), roach (*L. rutilus*) and perch (*Perca fluviatilis*) are also used successfully as a winter food for pigs. A small area of southern Sweden gave an average annual catch of 154,000 pounds of these three species—enough to provide the full requirement of animal protein for the production of some 880,000 pounds of bacon (pork). A weak stock of pigs reacted positively to the fish diet, showing considerable improvement in growth and health. The only difficulty is the problem of distributing the fish while still fresh. Although fish are quantitatively important in many lakes, they usually represent only a fraction of the vegetal productivity of the same lakes. For example, the fishery yield in a lake in southern Sweden represents only 0.1 percent of the primary (= vegetal) productivity.

Angling is one of the most popular activities for outdoor recreation. In North America, it is estimated, about twenty million anglers are annually attracted to freshwater lakes and streams. In Sweden, a country with seven million inhabitants, there are about two million anglers who pursue their hobby for fun six to twenty "fishing days" per person per year. This means about twenty million recreation days dedicated to sport fishing.

Many tropical freshwater areas are very productive and have fish, birds, and mammals, including man, at the end of the food chains. In addition to a tremendous variety of fish species, which utilize different food niches and habitats, the quantities of fish are great, supporting important fisheries, and providing food for large human populations. Lake Edward in Africa has been mentioned. Other African lakes that supply large amounts of fish provender include Lakes Tanganyika, Albert, and Chad. The latter yields about 100,000 tons of fish annually. In Uganda, Lake George produces an annual fish crop of 35 tons per square mile, and Lake Nakavali, where *Tilapia* has been introduced, 60 tons. High protein harvests are also obtained from fish ponds, to which fertilizer and fish food are often added. Fish ponds in southeastern Asia yield 13,500 pounds per acre. In China, in the Kwantung Province alone, are some 200,000 acres of fish ponds that yield annually a total of 350,000 tons of fish. African ponds stocked with *Tilapia* of various species are extremely productive. Central European fish ponds stocked with carp (*Cyprinus carpio*) produce up to 1,400 pounds per acre a year.

However, pond culture for producing food fish is an artificial rearing of various species that in natural waters once produced larger quantities, without any investment and with much lower labor costs. A restoration of natural water productivity on a large scale would be of immense benefit to the human community not only in the region concerned but also on a national level. The return per acre from aquatic vegetation, fish, wildfowl, fur mammals, and recreational use would equal or even be larger than what many farms yield. At the same time, the nearby landscape and its soils would regain health, thanks to the reestablished water cycle.

Ducks and geese are produced mainly in temperate wetlands. The most productive wildfowl habitats in the world are the North American potholes, which are filled by surface drainage from their immediate surroundings. In general they lose water mainly by evaporation. Marshes and lakes occur in the same areas. These wetlands not only provide extensive habitats for breeding but also receive concentrations of wildfowl engaged in moulting, resting, feeding, and migrating.

The abundance of some algae (*Spirulina maxima*) in many African lakes provides a useful source of food. These algae can be easily collected and would serve man and domestic animals.

Drainage and Irrigation

Drainage of lakes, marshes, and streams during the last 150 years has destroyed enormous areas of productive wetlands and surrounding regions. North America and Europe, in particular, have been drastically reshaped by this means in a very short time. Of 127 million acres of wetlands existing in the United States in Colonial times, more than 100 million acres have been drained. Immense areas that once grew cattails, wild rice, and pondweeds and produced wildfowl in great numbers now support agricultural crops, factories, villages, towns, and airports. What this giant transformation means for the water economy of a continent is obvious to those who are aware of the intimate relationship between water resources and the long-term productivity of the soil. The lakes and marshes in the prairie provinces of Canada have met the same fate. In Europe, marshes and bogs have been altered more during the past 150 years than have any other type of landscape. So many of them have been drained that the total area of permanent marshland and of land that is flooded annually has shrunk to a mere fraction of its former size.

The extensive and costly drainage projects have not always accomplished their purpose. Many in fact have been total economic failures. Instead of providing the productive agricultural land anticipated, they have resulted only in a few poor pastures and in large areas that are virtually useless. Vendée, France, is an example of an area that has suffered gravely in the aftermath of a drainage project. Great areas were drained there, but cultivation was found to be unprofitable and ceased after a few years. In some parts no attempt was made even to start cultivating the land. The elimination and subsequent abandonment of nature's resources by drainage still continues. When Normandy's great Marais Vernier was drained, it was predicted that the region would be transformed into a granary, a "Ukraine in Normandy," but nothing worthwhile came of it. Marais Vernier was much more productive in its original state than after it was drained and plowed.

There is a frighteningly long list of former lake and marsh areas throughout Europe and North America that have been destroyed unnecessarily through drainage and ignorance of ecological realities. Agriculturists, water engineers, and politicians are thus guilty on a criminal level of ruining productive wetlands where such projects were undertaken in large part at public expense. This is a form of farm subsidy that has no regard for national prosperity. Conversion to agriculture is seldom the wisest or the most economical way of utilizing a wetland resource. The benefit of many drainage schemes has been to the promoters, but hardly to anybody else. Shortsighted destruction of natural resources that only serve crudely selfish purposes must be called scandalous, and in the public interest should be so acknowledged. In recent years a few governments have finally realized that the drainage of remaining wetlands must be discouraged, and that many lakes and marshes already drained ought to be restored.

Tragically, a high percentage of the drainage failures have included the most productive wetlands, representing an array of values and a very high productivity of fish and wildfowl. The most negative economic consequence of drainage has been the sinking of the groundwater table, adversely affecting large surrounding, previously fertile areas and promoting soil erosion.

A drainage project, like all other development plans involving land use, cannot be of public benefit if it destroys existing values greater than those it intends to provide. Careful studies by unbiased experts, including ecologists, should always precede decisions to drain land or to provide financial assistance to such projects.

Irrigation of arid lands and deserts is often a means of recapturing previously fertile areas that man has destroyed. Although such operations may be successful, they often create new problems, because water running from its sources to the sea has many functions along its course. Each catchment area must be considered as a hydrographic unity, and interference with one function within the unity will inevitably lead to consequences in another.

One of the most successful irrigation schemes of modern time is in Israel, where a new landscape has been created from the desert, providing not only crops but also an environment where man can find variety and into which plants and animals will settle automatically in new-found habitats. However, it remains to be seen whether or not this creation of fertile lands by irrigation will have long-run adverse effects elsewhere within the drainage basin. Restoration of the former landscape is one procedure that takes time but will surely pay. Irrigation through artificial canal systems, on the other hand, may have rapid initial success but later on may prove environmentally fatal, particularly in a subtropical or tropical landscape. A great danger exists when irrigation structures bring entire river basins into agricultural production. A reaction chain of long-term negative effects is often released, involving the disappearance of fertilizing silts, increased salinization, less aeration of soils through seasonal dryings, and exhaustion of the lands despite increasing amounts of imported chemical fertilizers, including pesticides to combat crop parasites, which are favored by monocultures. As noted, natural irrigation through seasonal floodings has kept the Nile valley of Egypt fertile on an optimal level during the longest, continuous agricultural period known—a happy historical condition that may be brought to an end by the new artificial irrigating system based on the Aswan Dam.

History is full of irrigation failures in subtropical countries. In northern India more than 8 million acres of arable land have been destroyed by the progression of salinity and by secondary formations as a result of irrigation. Millions of acres in the Indus Valley of West Pakistan and hundreds of thousands in Central Asia, Iraq, Iran, and Turkey have met the same fate. Soil deterioration through irrigation is particularly pronounced where natural drainage is poor, rainfall is low, and evaporation is high. Irrigation upsets the balance between infiltration, capillarity, and evaporation. The result is an accumulation of salts in the soil and even in the groundwater table, which may concentrate in such high amounts that it reaches toxic levels.

One of the most dramatic irrigation consequences is what has happened to the Caspian Sea, largest inland body of water in the world. From 1896 to 1929 the water level decreased by about one and a half feet; between 1929 and 1946 there was a further decrease of more than six feet. Since 1946 the rate of decrease has accelerated. The level of the Caspian Sea has dropped about nine feet during the last twenty-five years. The most important factor contributing to this tremendous change is the reduced flow from the Volga and other rivers. As much as 76 percent of the water that once flowed into the Caspian Sea came from the Volga, but now about 50 percent of the river water never reaches the sea because it has been harnessed for irrigation and other consumptional uses. The impact of such a drastic environmental change is tremendous.

The lower Volga, the Volga delta, and the Caspian Sea itself as well as the surrounding land have been harmfully affected and have lost much of their previously high productivity. The Caspian region is virtually collapsing. "On the whole, one gets a strong impression that an extremely rich complex of living natural resources has been exploited so intensively by man that the area is now turning against him and his interests" (Curry-Lindahl, 1964). Since I wrote this passage, the situation in the U.S.S.R. Caspian region has become worse, and chiefly it is due to irrigation.

The United States also has examples of landscapes altered adversely by irrigation. Death Valley and Imperial Valley in California and parts of New Mexico are infested by salinity induced by irrigation.

Natural irrigation is seasonal, while artificial irrigation is perennial. The latter factor favours greatly the vectors of diseases which are fatal to man, such as bilharzia and malaria, because snails and mosquitoes transmitting these diseases thrive in the stagnant water.

Irrigation schemes have in some areas, like drainage schemes in others, drastically lowered the groundwater table with fatal environmental effects. In northern California, the groundwater level in an area of 100 by 300 miles has been lowered by as much as 200 feet. As the groundwater was drawn out, the ground subsided and produced serious disruptions of human activities on the land surface.

Irrigation projects in developing countries, induced by international or foreign agencies for technical and economic assistance, have led to waterlogging and salination of soils, resulting in a net loss of arable lands.

In California, scarcely a major river has avoided being dammed, with the result that most of the drainage from the southern mountains now flows through concrete canals. The existence of vegetation, wildlife, and human beings along the transformed natural water courses is threatened because vast quantities of water is artificially moved from one part of the country to another. When the vegetation cover in an area is gone, its water runoff will be considerably less, or will disappear entirely if there is a temporary local increase of water yield through lessened vegetation transpiration. Hence artificial irrigation schemes may initially favor a particular region, but in the long run adversely affect the drainage basin as a unit and make its future uncertain. Technology may solve the problems of water shortage, but basically this is an ecological and biological dilemma, for the world's human population of today exceeds the carrying capacity of its environment. We destroy more than we build. Man's faith in modern technology is dangerous to him, and with regard to the renewable natural resources, up to now it has not proven justifiable.

Water Power versus Living Waters

Potential energy power is available along streams and rivers. Man first utilized this power by installing waterwheels. But it was not until the electrification of industries, railways, and almost everything serving domestic needs that harnessing of water power was undertaken to such an extent as to modify the countryside more than any other human activity. In less than fifty years hydroelectric schemes on all continents except Antarctica have reshaped landscapes so greatly that the results can only be compared to what Ice Age glaciations of the northern hemisphere achieved during thousands of years. Man has become a geologic force, but not only for the good.

The conservation movement in North America, Europe, and Australia has fought bitterly for decades to save at least some of the river systems and mountainous countries from exploitation for water power, which means destruction forever of numerous other natural resources. Obviously a modern country must produce electricity, and water power is a renewable, nondiminishing resource that can and should be utilized for the purpose. Conservationists fully agree on this, but they react strongly against the attitude of politicians, economists, engineers, and others that immediate and full use should be made of all potential water power possibilities. That philosophy is dangerously shortsighted, indeed blind to other water resource values. It is even so technocratically narrow that it fails to comprehend that hydroelectric exploitation of all available rivers suitable for the purpose is uneconomic for a country.

Regulation of rivers and lakes for hydroelectricity must be planned in relation to the far-reaching negative consequences that also ensue. These are manifold. Salmon and other economically important fish disappear. Fishery yields decrease not only in the river system concerned, but also at sea. The natural productivity of rivers and lakes collapses, chiefly due to constant fluctuations in water level. The shores of dammed rivers and lakes are destroyed. Wave action and ice lifting on beaches cause serious erosion. Surrounding meadows and forests covering immense areas are drowned, eliminating grazing, forestry, and other uses. Terrestrial wildlife is wiped out. The microclimates of the soil, and sometimes the entire local climate, are affected. The whole ecosystem with all its productive habitats is smashed. Recreational activities are lost. The beauty of the landscape is replaced by constructions having a mighty concrete sterility, aesthetically appealing only to a few. In short, the living rivers with their turbulent rapids and thundering waterfalls are tamed to silence, imprisoned by gigantic dam walls and turbines through which the water is allowed to pass only when an engineer pushes a button. For a living landscape teeming with organisms, a dead one is substituted.

It is true that hydroelectric development may not always deprive the salmon of their spawning grounds. Artificial fishways and reproduction may compensate part of the losses. Other measures may reduce the environmental damage. In the United States, in particular, construction of enormous dams and reservoirs for hydroelectric power and irrigation is claimed to confer other benefits in addition to its primary purpose. These include flood control, fish and wildlife enhancement, and increased recreation possibilities. Flood control is in fact an important and usually favorable result of dam constructions in highlands, but so far as fish and wildlife and recreation are concerned, with local exceptions, it is an overstatement to claim that they are generally benefited—particularly when the claim is tempered by a future perspective.

Many dams, within a relatively short time after completion, fill up with sediments that reduce their capacity and necessitate, as man sees it, the building of new dams in the same river system. Speaking of the situation in California in 1965, Raymond F. Dassmann questioned the wisdom of the water-planning schemes: "Inevitably an end must be reached when there are no more rivers to dam, and no more water surplus to be shifted. When this time comes, either growth must cease or we must find some new way of providing water that differs from the dam and canal system we now use. . . . Will southern California truly be a better place to live when it has 20,000,000 people instead of 10,000,000? Will we really be better off when we have paved over all of the orange groves of Los Angeles, and developed new ones in the desert?"

This problem does not just face the state of California. It concerns a great part of the world. If every river system of a country is destroyed, there is nothing left as a basis for comparison when man needs scientific answers and guidelines to management problems. This could be economically fatal. Moreover, our generation has a duty to keep intact at least a few water systems in each country for the generations of tomorrow. They will have the right to see for themselves what a living, undamaged river of their country is like along its full course from source to the sea.

This viewpoint is important for such areas as the Grand Canyon in Arizona and the Murchison Falls of the Nile in Uganda, which have both recently been suggested as sites for big dams or electric power plant stations. These projects would not just destroy some of the most scientifically significant areas of the regions concerned; they would also deprive the world of two of its wonders, making us all much poorer. The United States and Uganda are custodians of these marvels and are therefore morally obliged to their own people and the world to preserve them.

Fortunately, an electric power project associated with the proposed giant Rampart Dam in Alaska was carefully investigated by ecologists and other concerned interests over a period of several years before a final decision was taken. As a result, it was found that the vast area under threat of destruction would be economically more desirable if kept intact than if it were developed for electric power.

What can happen to natural resources smashed by thoughtless exploitation for electric power is shown by the crisis in the great Columbia River basin of the United States' Pacific Northwest. There, economically and recreationally very important natural resources like the runs of salmon, seagoing trout, and shad (*Alosa sapidissima*) in both the Snake and Columbia rivers are being made to pay the price of progress. In 1968 more than 50 percent of the summer chinook salmon (*Oncorhynchus tshawytscha*) failed to appear at their traditional grounds of the Snake River and the upper Columbia due to serious passage problems over dams in the Columbia River. The summer run of steelhead trout (*Salmo gairdneri*—rainbow trout that migrate to sea) appeared to be the smallest on record. The run of the sockeye salmon (*Oncorhynchus nerka*) in the Columbia River has been seriously affected. Even the shad, which has recently extended its range far into the upper reaches of the Columbia River, now have difficulty in making passage. The direct reasons for these fatal consequences to the upriver runs are cumulative dam-to-dam losses to both upstream and downstream migrants, virtually complete transformation of the natural river environments, and that water spilling over the dams becomes saturated with nitrogen which reaches lethal levels in some areas.

Biologists of the Oregon Fish Commission have stressed that it should be painfully evident to all concerned that the hydroelectric exploitation of the Columbia River system has reached such a level of environmental destruction that it threatens to wipe out the great, historic, extremely valuable fishery based on the anadromous fish runs of the Columbia River. Is it "progress" to destroy a renewable natural resource of this magnitude by developing another, which yields no proteins?

Many rivers, valleys, and plains in various parts of the world have been transformed into man-made lakes. Many of these artificial lakes are enormous and have modified entire ecosystems of large regions. Tremendous environmental disturbance results from such alterations.

Besides producing electric energy, man-made lakes are also used for flood control, water storage, irrigation, fish production, and recreation. All this is on the credit side, but the balance is not always positive. Unexpected repercussions become clearly visible after some years, even before the dam has been filled. Nevertheless, most economists claim there is no doubt that the benefits brought about by man-made lakes outweigh the problems created. This is probably true in some areas, but on the other hand, the negative environmental results could have been foreseen, greatly reduced, or perhaps even avoided had ecologists been consulted at an early stage. In tropical and subtropical areas dams seem to be less successful.

It is often claimed that increase in fish yield will automatically be one of the blessings the region concerned will receive from artificial dams. Usually man-made lakes respond to these expectations in their initial stage, when the water rises, conquering new areas with an abundance of food organisms favouring the fish. This situation prevails as long as the dammed river floods dry land and for some years thereafter. The fish increase during this stage, but when the former terrestrial habitats have been killed, impoverished, or stabilized by the dammed lake, there is no longer an abundance of food. The fish decrease and the catches drop far below the estimated yield. The artificial fluctuations of the water level in dams do not favour living organisms and therefore eliminate food chains.

Pollution

Dirty water is repellent and disgusting to man, yet he is the only creature on earth that pollutes water. He does it deliberately on a gigantic scale and at an increasing rate, to such an extent that the pollution has become a serious threat to human health, indirectly and directly. Enormous quantities of waste from urban communities, industries, and agriculture are daily discharged into the water, which has great self-cleaning ability, but not without limit. A fearfully high percentage of the discharges is poisonous. That the pollution has gone on for decades on an even worse scale without substantial reactions by man himself seems to be explained by the fact that he does not see directly what is going on below the surface of the water. Apparently municipalities and industries believed that waterways could indefinitely absorb, assimilate, dilute, and carry away astronomic quantities of wastes. This philosophy was wrong, and will cost the taxpayers tremendous sums for many years ahead, because it is now necessary to pay the bill of a debt that has steadily accumulated for a century. Much of the pollution is direct, but to a great extent it is also indirect, through airborne fallout. Water pollution is worst in countries of advanced development. In the matter of installations for water treatment systems, they are very backward. Future generations will condemn this environmental destruction as criminal—but perhaps they will forgive us, since by then it will be recognized that twentieth century mankind was ecologically illiterate.

In developed countries the cost of clean water is tremendous. In the United States alone the cost of collecting and adequately treating municipal and industrial wastes discharged in the waterways during the period 1969–73 may range from $26 to $29 billion, according to a report prepared by the U. S. Department of the Interior.

The use of pesticides in agriculture and forestry, as well as the use of mercury compounds by many industries, has not only polluted the waters but poisoned them as well. Pesticides enter water in several ways, but chiefly through runoff from treated areas, waste discharge by producers and industries, or directly from aerial spraying. In Sweden a great number of lakes, rivers, and marine archipelagoes have been blacklisted as contaminated by mercury to such a degree that the fish produced there are too poisonous to be eaten. The fatal effects on the populations of fish-eating predators like grebes, white-tailed eagles, ospreys, and otters are obvious. These species constitute the same final link in the food chain as man. Seventeen years after the conservationists had warned the Swedish government of the danger of using pesticides indiscriminately, measures were finally taken to ban commercial fishing in many lakes. The fish in these lakes were found to contain such a high percentage of mercury residues in their flesh that as food they were judged a danger to human health and life. That food so contaminated is poisonous is more than just a theory. Tragic experience in Japan has conclusively demonstrated the mortal effects on human beings of eating mercury-contaminated flesh. And the animals provide a long series of data. For example, the populations of the otter and many species of birds of prey in Sweden have fallen to a near-extinction level during one decade, due to mercury poisoning. Moreover, the long-term genetical effects of mercury intake are not yet clear. Sterility has occurred in birds, and embryonic abnormalities are not excluded.

Also in North American lakes mercury residues in edible fish exceed the tolerance level (0.5 ppm mercury in the wet edible tissue of fish) presently in use by the U. S. Food and Drug Administration and the Canadian Food and Drug Directorate. Examples from Lakes St. Clair and Erie in the United States and Canada are significant. In Lake St. Clair fish have been found generally to exceed the level mentioned above regardless of species and location in the lake. The level of mercury in wall-eyed pike has been known to reach 5 ppm; while in other fish of a less predatory nature, e.g., perch, levels as high as 2 ppm have been recorded. Consequently, commercial and sport fishing has been prohibited on both the U. S. and Canadian sides of Lake St. Clair. In the Western Basin of Lake Erie where mercury contamination is lower, wall-eyed pike have the highest levels, averaging somewhat over the 0.5 limit and peaking at around 2 ppm. Perch seem to be straddling the action level with most specimens that have been tested below. The other species of importance, smelts, has been running well below the action level, generally from about 0.05 to 0.2 ppm. In the Western Basin, Canada has placed an embargo on wall-eye and perch, i.e., they cannot be sold until analyzed and cleared by the Canadian authorities. As of April 29, 1970, the state of Michigan announced a similar embargo in the areas of Lake Erie under its jurisdiction. In the Ohio waters of Lake Erie, an embargo has been placed only on wall-eyed pike at present.

In June, 1971, the governments of Canada and the United States reached an agreement on a wide range of joint programs to restore Lakes Erie and Ontario by 1975. This is one of the most important conservation actions undertaken in the world and might have most significant results.

In 1970, mercury compounds were found in fish and in rivers

and lakes of not less than seventeen of the United States due to dumping by industries using mercury compounds in their processes. Recent findings (January, 1971) in the state of New York show that forty-three-year-old preserved fish have levels of mercury more than twice as high as the amount recently regarded as safe for human consumption. It remains to be learned whether this high mercury content reflects a high degree of pollution several decades ago or has other explanations. Possibly large amounts of mercury have been ingested over the last forty years by human beings feeding on fish of the New York lakes.

Coal and crude oil are recently reported to contain mercury that is released into the atmosphere when these fuels are burned. This will probably result in air-borne mercuric fallout in rain and dust. If this fallout were a major source of mercury contamination of water it would explain the contaminated fish from isolated ponds in the Adirondacks and also would imply that mercury pollution has been growing since the onset of the industrial revolution in the same way the level of the atmospheric lead pollution has been shown to correlate with the industrial revolution (Dickson, 1971). However, mercury data in birds from Sweden during the last century contradict this hypothesis.

Concentrations of DDT and dieldrin have been recorded in waterways in almost all industrial countries, particularly in the United States, where they are most critical in the Great Lakes. It was recently found (Poff and Degurse, 1970) that, in general, all mature trout and salmon from Lake Michigan contained residues in excess of the 5 mg/kg "action level" established by the U. S. Food and Drug Administration. Data from fish in other Great Lakes also show high levels of DDT as follows: Lake Superior up to 7.77 mg/kg, Lake Huron up to 6.90 mg/kg, Lake Erie up to 1.89 mg/kg, and Lake Ontario up to 4.32 mg/kg (Carr and Reinert, 1968). Similar data exist from other drainage basins in North America.

Recent investigation in Lake George, New York, has shown that DDT completely inhibited reproduction in lake trout. Although the egg contained relatively small amounts of DDT, the fry were killed at the time of final yolk sac absorption (Mrak, 1969). This is a parallel to adult fish which can carry DDT residues in their tissues and remain in good health until, by a shortage of food, they begin to use up their own stored fat, which poisons them with lethal effect.

The reason why fish or cold-blooded vertebrates, in general, living in poisoned waters are so dangerous to their warm-blooded predators—birds and mammals, including man—is that they seem to tolerate ingested or absorbed poisons to such an extent that they can store quantities of residue not lethal to them but which may kill or genetically affect their predators.

Mercury is but one offending element. Literally thousands of other chemical compounds that pollute the water bring death and devastation to previously unspoiled rivers and lakes. Aquatic organisms quickly absorb poisonous chemicals spilled or deliberately put in their element. Even if the poisoning hits just one species or group of species, the negative effect is usually multiplied in the habitat or the ecosystem. The destruction of algae by sewage wastes robs fish of life-giving oxygen. Fish also have a high direct vulnerability to pollution. In fact, in the United States pollution kills more fish every year than public conservation agencies can replace. Despite incomplete coverage, reports show that 15,235,000 fish were killed in 1968 alone, an increase of 31 percent over 1967. In 1969 the figure was an estimated 41,000,000 fish, an increase of more than 100 percent. The census on pollution fish kills in the United States was started in 1960. A total of 144,380,000 fish were reported killed through 1969 with a total of 3,200 incidents. Two-thirds of the fish killed were of commercial value. The biggest single kill was in Lake Thonotosassa at Plant City, Florida, where the accumulated industrial effluent and municipal sewage reduced the oxygen content of the water so that 26,500,000 fish died. Pesticides killed more than 5 million fish in the Lower Mississippi River and over 2 million fish in a Florida marsh. In 1965 I witnessed a tremendous mass death of fish in Lake Balaton, Hungary's most important fish protein resource, caused by agricultural pesticides. On May 5, 1970, an estimated 349,000 fish died in seven miles of Crooked Creek, Missouri, following the dumping of an unknown quantity of toxic material (a solution of 10 percent chlordane and 5 percent of malathion in xylene). In addition to the fish all other aquatic life of the stream was killed for a distance of two miles downstream from the dumping. The same polluted waters that killed fish are used and reused in many ways by man. Can humans survive where animals die?

The majority of fish kills in the United States are caused by toxic wastes from industrial operations and pesticides. In 1969, over 70 percent of the kills were caused by industrial pollution and 20 percent by agricultural pollution. Most of the fish killed were species already tolerant of water pollution. The sensitive fish had already been wiped out.

This was the case for the Rhine in Europe, when in 1969 this river was exposed to a sudden industrial exhaust in Germany, which polluted a huge part of the river with highly toxic chemicals. The effect was a mass kill of fish and other aquatic organisms along the Rhine in Germany and the Netherlands as well as destruction of Dutch water supplies for several communities. The source of this pollution in Germany was not detected soon enough. The Netherlands was hit before the German authorities had time to warn of the lethally poisoned water flowing down the Rhine.

In 1971 the contamination of the Rhine from heavy metals and pesticides has reached alarming proportions. About 80 tons of arsenic, 300 tons of copper, 20 tons of cadmium, 10 tons of mercury, and 900 kilograms of insecticides enter the Netherlands each month via the water of the Rhine. The river discharges 60 million tons of dissolved solids into the North Sea every year.

In Finland 50 percent of the population is affected by seriously polluted water.

Many lakes in tropical and subtropical regions have no outlets to the sea and function as closed or internal drainage basins. Such lakes are often very productive and have a rich fauna. Many of them are alkaline. If toxic chemicals are used in industry and agriculture within such closed water systems, these residues certainly end up in the lakes, where they and other pollutants accumulate. Several such internal drainage basins are very large and play an important role in the ecology and economy of regions which cover several countries. Lake Chad in Africa is such a closed drainage basin where four countries share the ownership. The Okawango Swamps and the Makarikari Depression is another closed water system covering a large area in Angola and Botswana.

A similar situation as far as pollution is concerned is found in many lakes and swamp areas, particularly in the tropics and subtropics, where the outflowing river carries less water than the inflowing river. The explanation of this phenomenon is that the balance is made up by the difference between precipitation and evaporation. But unfortunately many biocides do not evaporate, they just accumulate.

The toxicity levels of water pollution in industrial countries seem to be rising due to increasingly complex manufacturing processes while agricultural toxic wastes increase in volume. Mercury pollution is widespread but still undetected in many countries, possibly existing at the same high level as in the United States and Sweden. Many governments do not seem to be aware of this toxic danger to aquatic and terrestrial life, including humans. In addition, current levels of lead, cadmium, and other poisonous substances masked in the water certainly have effects on animals and man, but we still do not know why.

The nervous system of both freshwater and marine fish is affected by pesticides, which cause respiratory and balance difficulties, apathy and death. DDT reduces the ability of salmon to detect temperature changes and, therefore, they fail to find their ancestral spawning grounds.

The disposal of atomic wastes is a major problem of extreme importance. Even if atomic power stations do not dump their waste into inland waters, radioactive raindrops cannot be avoided. They disturb the water supplies and growing crops and other vegetation through which they pass to cow's milk that is in turn consumed by humans.

Thermal pollution caused by nuclear power plants will certainly become a conservation problem of increasing scope, not only in freshwater but also in marine areas. Radical changes of thermal-nutrient structural relationships in lakes, rivers, estuaries, and coastal waters will inevitably have serious impacts on the aquatic environments. These changes may produce various undesirable effects on the local climate, oxygen circulation, vegetation growth, fisheries, and recreation. Many such effects are already well documented in various United States lakes. Proliferation of single-celled algae in warmed water may lead to development of substances that are toxic to vertebrates, or by depleting the oxygen supply may destroy aquatic habitats, including important food for fish. Small amounts of radioactive material may be expected to be released continuously with the water discharge.

Several technical means exist for cooling the heated water before it is discharged into a lake, river, or the sea. At present the most satisfactory method seems to be the use of evaporative towers with a closed-circulation cooling system. This method will be used in Vermont to control the thermal output to the Connecticut River. Permits for construction of atomic power plants should not be granted unless efficient cooling systems are a part of the project.

Of the many kinds of water pollution, one of the worst is oil pollution. We have already dealt with it in the chapter on the sea. Oil pollution also occurs in fresh water. As in the sea, the terrible effects on oil-soaked birds are those chiefly observed, while subwater damages are less observable. In March, 1960—to cite just one example—oil pollution of the Detroit River killed twelve thousand ducks, mostly canvasbacks (*Aythya valisineria*).

Outboard motors contribute to water pollution at a much greater rate than has been realized. Ten to forty percent of the outboard's fuel is often discharged unburned directly into the water. Most case studies have shown that oil has an adverse effect on fish growth and longevity. Unfortunately, fish tend to be attracted to gasoline-polluted water, only to be harmed by it.

In Lake George, located in the Adirondack Mountains of New York State, a minimum of forty thousand gallons of outboard fuel was discharged into the lake during the 1967 summer season. Since only about 10 percent of the volume of the lake is renewed annually, a considerable portion of the fuel is left to contaminate the aquatic life. If the fuel pollution of Lake George increases, which is very likely due to increasing human population, the residual pollutant level will rise. Smell and taste of gasoline and oil will become noticeable in fish as well as in water supplies.

Lake George is just one example. Thousands of other navigable waters of the United States alone are polluted annually by at least 100 million gallons of unburned outboard fuel. Since unburned fuel does not seem to undergo decomposition, it is gradually increasing the pollution of freshwater lakes, along with oily mixtures. This accumulation of gasoline and oil can take years to reach a fatal level, with serious effects on aquatic life and water quality. Public opinion must press outboard motor manufacturers to improve their products, which today constitute an environmental menace.

Deliberate polluting of fresh water with poison designed to destroy natural growths also kills off animal life and renders the aquatic habitat unfit for useful animals like wildfowl and mammals. Other methods exist to get rid of unwanted vegetation without destroying everything living in the water treated.

An interesting example of "biological pollution" comes from observations from the Ardennes in Belgium, where spruce planted too close to streams have contaminated the waters, preventing trout from spawning.

Though we do not know enough about the extent, the effects, and the control of water pollution, we know more about it than we put to practical use. While it is possible that future water treatments and other actions against pollution may differ in various major river systems, there are basic measures and minimum management obligations that should be made a part of national legislation. Water quality criteria must be established and stressed in a national water conservation policy.

Aquatic Plant and Animal Life

Nutrient salts occur in fresh waters, their amounts varying largely from one lake or marsh to another, and also seasonally. In general, these substances together with other components govern aquatic plant growth, which in its turn is a limiting factor controlling aquatic animal populations. Organisms show specialized adaptations for particular niches in the freshwater habitats from the bottom to the surface. Therefore it is important to consider all aspects of organic life produced in the ecosystem of a lake, a marsh, or a river before making decisions involving exploitation or habitat modification. Conservation and management of aquatic biocommunities are extremely complex procedures. Up to now they have failed to prevent various measures that have proved ineffective at best and sometimes have been disastrous.

Based on the productivity of algae and other green plants, the bulk of animal biomass in freshwater habitats is mostly built up by insects, crustaceans, and fish. To these food chains come other consumers such as birds and mammals that must also be considered in an accounting of the biological productivity of a lake or a marsh, though many terrestrial vertebrates, like some fish species, utilize the area only a part of the year. Also a great number of terrestrial plant and animal species utilize the shores of a lake, marsh, or river, and the littoral zone community produces much that is not usually included in estimates of freshwater productivity. The rich plant and animal life of Lake Baikal in Siberia, the deepest lake in the world (5,658 feet), is concentrated in the littoral zone. The lake is now threatened by hydroelectric power installations and regulation of its water. If the level of Lake Baikal is reduced by only ten to fifteen feet, the plants and animals along the shores will be severely damaged.

A few words should be said about the introduction of exotic aquatic plants into a freshwater environment. This may lead to catastrophic consequences, not only for the biocommunities concerned but also for the boat traffic in rivers and lakes. Introducing the American water hyacinth (*Eichhornia crassipes*) in the Old World has, for example, caused tremendous economic losses and much difficulty. Outside its original habitats, this plant spreads in spectacular fashion with great rapidity over wide water areas, which it covers with a green blanket.

Eutrophication

In temperate regions most lakes are either eutrophic (rich in nutrients and vegetation) or oligotrophic (poor in nutrients and vegetation). Lakes which can be characterized as intermediate between these two stages are mesotrophic. Eutrophication is a natural process in many lakes and marshes, particularly in previously glaciated areas where the land upheaval makes water bodies shallower and gradually more eutrophic until they die as lakes or marshes to become bogs, fens, meadows, or cultivated land.

However, eutrophication can also be a consequence of man-made activities, chiefly due to the introduction of excess nutrients in lakes and other types of water. The artificial nutrient enrichment in lakes through organic waste loads accelerates the eutrophication process tremendously, to such an extent that it at present constitutes one of the most serious water pollution problems. The water quality deteriorates rapidly, and many lakes are already useless as reservoirs for potable water or cannot even be used for recreation (bathing) not to speak of the degradation or even elimination of fish resources.

Fish populations are good indices of eutrophication as well as changes of the species composition of bacterial, planktonic, and benthic communities. Species of fish react differently to various degrees of eutrophication pollution, chiefly because when excessive aquatic vegetation decomposes it consumes the deep-water oxygen so vital for fish and other aquatic life organisms. Catastrophic kills of fish often occur in eutrophically polluted, temperate lakes when the latter are ice covered for long periods. Moreover, eutrophication pollution causes "smelling lakes," when the decomposition odors from excess aquatic plants fill the water and foul the air. Eutrophication pollution also chokes the open water and makes fishery activities and communications difficult. The latter effect can also be caused by introduction of exotic plant species. Eutrophication pollution also strikes estuaries and slow-flowing streams.

The direct cause of accelerating eutrophication through pollution seems to be phosphates which stimulate particularly algal growth and also that of larger aquatic plants. Carbon and nitrogen also contribute to eutrophication, as well as other substances besides phosphates and nitrates, which are synthesized in the biological treatment of sewage.

Both in North America and Europe the largest amount of phosphates and nitrogen reaches the lakes from municipal waste discharges and through runoff of phosphate fertilizers and other manure from agricultural land. It must be stressed, however, that there are also beneficial aspects of eutrophication of lakes and ponds. This is mostly in the case of natural eutrophication and in artificial fish ponds. In the latter, fish production of desirable species may increase greatly by artificial fertilizing with sewage effluent or other substances.

Fishes and Fisheries

The protein value of fish inhabiting fresh water is immense, though much inferior to what the seas produce. Regrettably, man has destroyed enormous areas of productive freshwater areas. He has so mistreated the rivers that many of them, particularly in Europe and North America, have been killed in a biological sense. In the past, migratory fish like salmons often gave spectacular evidence of their abundance. From the nineteenth century come many accounts of the former richness of rivers. In the Columbia and Sacramento rivers of North America, upstream migration of Pacific salmons (*Oncorhynchus*) lasted from spring to autumn. The mass of migrating fish was so enormous that nets could not be used. Instead, large bucket wheels were used to scoop them out. One such wheel could catch as many as fourteen thousand salmon in one day. In the relatively short rivers in eastern Siberia another salmon, the keta (*O. keta*), swam upstream in such great numbers the streams were not large enough to hold the fish. Their backs protruded out of the water, and many were crowded against the river banks. Probably none of these salmon ever returned to the sea, but those that were able to spawn successfully provided through their offspring vast new upriver migrations some years later. So it went every year, but such magnificent concentrations scarcely exist any longer.

In fact, eleven concrete dam walls across the Columbia River are standing between the salmons and their ancestral spawning grounds. This is too much for the salmon populations of this water system. The former runs of great salmon, steelhead, chinook, sockeye, and coho salmon are at present blocked off from thousands of miles of spawning area. Severe habitat loss in the Columbia River system has taken great toll of these salmon. More than half of the natural spawning areas of the Columbia Basin are at present denied to fish migrating from streams and rivers to the sea and back again. In addition to the disappearance of spawning grounds through inundation, the dams have added to the difficulties. Tens of thousands of salmon perish every year while trying in vain to find and climb the artificial fishways, which the engineers claim will solve all problems. To this come the even more serious losses of millions of young salmon migrating downstream to the Pacific. They are either sucked through the turbines and spewed out killed or mutilated, or they are worn and battered from going over the dam spillways and plunging into the turbulence below.

Despite the enormous decrease of salmon runs in the Columbia River, it is still the world's largest producer of chinook and steelhead salmon and the major contributor to the ocean salmon fishery in the Pacific from Alaska to California. This is primarily due to the very costly hatchery programs in Washington and Oregon, financed by federal money.

Likewise, the Atlantic salmon (*Salmo salar*) once almost overfilled the rivers of North America and Europe. As late as 1947, near Strasbourg in France, 143 salmon were caught in the Rhine River in one day. Today hardly a salmon puts its nose into the mouth of the Rhine. Most rivers of continental Europe that once had impressive runs of salmon are now deprived of this economically important fish, as well as trout and other salmonoids. In North America the Atlantic salmon has been exterminated in the rivers south of Maine. Everywhere the Atlantic salmon river fisheries have been reduced to a fraction of their former numbers and activity. While most catches of this salmon are made in the sea, this fishery is dependent on salmon productivity in upriver regions. Man's destruction of the freshwater breeding grounds of North Atlantic salmon during the last hundred years has reduced the total annual catch to about one-tenth its former volume, despite intensified deep-sea fishing efforts. The decline of the Atlantic salmon is a shocking example of wasteful destruction of a protein resource by pollution and by agricultural and industrial exploitation of the rivers that produce the resource. To this comes a new threat. In 1964 some of the most important marine feeding grounds for the Atlantic salmon were discovered off Greenland. Immediately industrial fisheries moved in. The commercial catch skyrocketed from 36 metric tons in 1965 to more than 1,000 in 1969.

Such heavy pressure in all waters where the Atlantic salmon occurs risks depleting through short-term greediness one of the world's most important marine and freshwater resources.

In 1969 delegates from eleven countries attended a meeting of the International Commission for Northwest Atlantic Fisheries (ICNAF) where a resolution was passed calling for a high-seas ban on salmon fishing in the Atlantic. Denmark and Germany voted against the resolution. At a 1970 meeting a total ban was again called for, and this time only Denmark voted against it. Negotiations resulted in a one-year compromise maintaining the 1971 catch at the 1969 level. The Danish trawlers have been trawling in the Atlantic off West Greenland with nylon nets since 1964 and their take has roughly doubled each year. In 1969 the Danish commercial fleet took some three million pounds. This ruthless exploitation of an international renewable resource is another example of how greed blinds man.

The runs of another important food fish, the gizzard shad (*Dorosoma cepedianum*), in streams along the North American East Coast are also gone, or have decreased to a shadow of their former glory. In the Great Lakes the fisheries not long ago were yielding the most valuable freshwater fish of North America, but the value of the Great Lakes fisheries has been reduced to almost nothing in comparison with the revenues they formerly earned.

In North America and Europe artificial salmon-raising is undertaken on a giant scale. This work is designed to compensate for the loss of natural spawning sites, which either have been destroyed or have been placed beyond reach of fish by the interposing of hydroelectric power stations in the rivers. Smolts—young salmon ready to undertake their first migration to the sea—are produced and fed artificially in ponds, and at the proper stage are released in the rivers below the dams and near the sea. Biologically it seems to be possible under favorable conditions to compensate artificially for the lost natural salmon reproduction. However, the costs are very high and the risks are great because human factors and failures are constantly involved. A high mortality of the unnatural concentrations of smolts in the ponds results from diseases that spread rapidly and cause catastrophes. As some diseases appear to be eliminated, new ones appear and the fight goes on. Furunculosis is a contagious disease with particularly virulent effects on salmonoid fish. Mortality may be 100 percent. This disease and others may eliminate a station's entire annual production of artificially raised smolts. This has happened in Sweden. Such losses do not occur in naturally reproducing salmon populations. Moreover, furunculosis has been transmitted through released salmon coming from artificial fish ponds to free-running water systems, where it has contaminated trout, salmon, and other salmonoid populations.

One may also question how many artifically raised smolts are necessary to maintain the salmon populations on a natural level, allowing sustained catches in the sea. And what is the genetic effect of continuous artificial salmon-raising with natural selection no longer fully operative? A degeneration of "population quality" may be a long-term consequence. In the case of salmon, man is manipulating a natural resource of high value without knowing just what the long-run effects will be. There are too many unknowns to permit claims that man can successfully replace the natural reproduction of salmon with an artificial scheme. Salmon are not cows. Even when artificially raised, they have to fend for themselves both in fresh water and marine habitats. Many complicated factors are involved, and migratory fish populations simply do not function naturally in the way man has arranged for them.

Fortunately, adult salmon return from the sea to the river where they spent their first years of life. Each stream has a unique scent, and the salmon can differentiate selectively among the myriads of characteristic odors. Professor Arthur D. Hasler of the University of Wisconsin has shown convincingly that migrating adult Pacific salmon are responding to the odor factor of a given stream by means of an imprinting process that occurs during the fingerling or smolt stages. Imprinting in the Atlantic salmon probably occurs when the smolts are in their second year. This learning process relative to environmental factors means that man can build a run of salmon up a selected river by introducing artificially hatched smolt before their imprinting stage. This has already been done in Canada, the United States, and Sweden, but it can be done on a much larger scale providing the available river systems are ecologically suitable for salmon. Here too management and restoration of freshwater habitats can help to restore the heavily decreasing salmon populations.

The salmon situation has been discussed at some length because from an economic standpoint salmon constitute one of the most important natural resources among fish reproduced in fresh water. However, running waters are not the only environment having economic importance because it favors fish life. Lakes and marshes as originally constituted, whether eutrophic or oligotrophic, were excellent fish habitats with annual harvests of substantial economic value.

Overfishing in many cases has brought about a decrease of fish populations and even the collapse of important fisheries, but for the most part human destruction of freshwater habitats explains why valuable food resources all over the world have been unnecessarily eliminated or seriously diminished. This is paradoxical because fish management measures have been in operation in both the northern and southern hemispheres for more than a century. Enormous annual investments have been and still are made to promote freshwater commercial fisheries and sports angling by artificial hatching and raising of various fish species for restocking purposes. Milliards of fry and fingerlings have been planted in lakes and rivers at tremendous costs, but in general these efforts have been useless or at least unproductive, because the fry mortality after restocking is very high. The losses are due to such various factors as unsuitable environment, too heavy competition for food as a result of overstocking, extreme vulnerability to predation because of overcrowding and lack of experience in the new habitats, and so forth. Stocking has introduced diseases into previously healthy waters, and other biological errors have been committed. Scientifically, it is regrettable that the mania to introduce fish has been implemented on such a large scale over vast territories of most continents without registering the original locality of the fish. These poorly planned operations have created a real hodgepodge of fish populations, races, and species in many lakes, where it is now impossible to find out what species or races originally existed there. Many species and even races show environmental modifications brought about by local conditions. This is especially true of various species of whitefish (*Coregonus*), but they have now been mixed to such an extent that it is difficult or impossible to trace the original forms, which would have been most suitable for introducing into particular lakes.

Normally, natural reproduction in suitable environments is far superior to artificial hatching and restocking. Therefore the management and, if necessary, restoration of aquatic habitats and fish populations are much more economical measures than artificial hatching and continuous restocking. As a rule the latter method pays only in waters where the fish populations have been exterminated, providing that the environment is appropriate for the species introduced. In streams and lakes near large cities, where the recreational fishing is so intensive that the fish have no chance to survive or reproduce, it is necessary to stock such waters continuously with fish during the angling season. If not, there would be nothing to fish. This is an entirely artificial situation.

What constitutes proper fish management does not seem to have been well understood until very recently, despite all the work done in the field. Man has always had difficulty in realizing that aquatic organisms and habitats function basically in the same way as does terrestrial life. Even today fish in American and Canadian national parks are treated and managed differently in principle from all other animals.

Suitable habitats, intraspecific competition, reproduction, and cropping are the fundamental divisions of freshwater fish management, but within those four major areas lie a wide diversity of subfactors. They are all interrelated. In many areas it is possible to obtain results by improving the environment for the fish already existing in a water. But often the local conditions have been upset by introduction of exotic fish, without careful study to determine if this step was really necessary in order to ameliorate the local conditions or increase the economic yield. It is true that exotic fish have often been introduced successfully in various countries and have led to a remarkable increase in protein production as well as in recreational angling opportunities. On the other hand, these actions also have negative results. Several fish species have been totally exterminated due to predation or competition from introduced exotic fish. In many waters, economically important fish have been reduced or have disappeared as a direct or indirect consequence of introducing exotic species.

The "cleaning" of lakes with rotenone, a poison that eliminates gill-breathing water animals, in order to destroy unwanted native species and provide free space for introducing exotic ones—mostly fish for sport—is a method that locally and exceptionally can be accepted for recreational purposes. But when such extermination is applied on a large and increasing scale over vast areas, involving many components of a water system, the long-term negative effects may be far-reaching. A large number of food chains are broken, both within and outside the water treated.

Improving human-altered habitats and restoring destroyed waters will reverse the steady decline of commercial freshwater fish. Management measures will contribute further in that direction. The simplest management tool in a healthy and productive habitat is to synchronize the fishing pressure to the crop produced. Both overfishing and underfishing should be avoided. The latter may result in overpopulation and increased intraspecific competition, often indicated by a stunted size of the species concerned. A diminishing average size of a fish population does not necessarily mean that it is overexploited; it may mean the contrary, particularly where there is a lack of natural predators.

Fishermen and sports anglers keep claiming that predators—fish-eating mammals, birds, and reptiles—are destructive to fish populations and the fishery interests. It is true that such predators eat fish, but this is not the same as being destructive, even if man is in the habit of catching the same kind of fish. Predation in natural habitats and biocommunities is a normal factor to which animal populations and often the whole ecosystem are adjusted. Investigations in all climatic regions on several continents have shown that predation is not destructive of natural fish populations and does not cause losses to the fishery interests. It must be remembered that very few fish-eating predators are specialized to just one species of fish. Usually their predation affects a rather wide array of species, of which many are also themselves predators on fish during various stages of their life. The proportion of important food and game fish in the diet of predatory mammals and birds is low in comparison with other food items consumed.

Predation damage may occur occasionally and locally in unnatural conditions—particularly where there are pronounced concentrations of artificially raised fish in ponds or just after releases of fry. Such conditions may attract fish-eating mammals and birds that, depending on local circumstances and species, may necessitate control measures. But costly control measures against predation on fish are, in general, bad investments. It would be much more useful to spend the same amount of money on habitat improvements such as reduction of pollution, restoration of drained areas, and the like.

Conservation of productive fish waters and a wise use of such resources are necessary to maintain the actual cropping yield at a sustained level. But this is not sufficient. There are many ways of increasing the harvest of fish populations in various types of fresh water. The most direct way to do so is to increase the fishing pressure in many lakes. Many fish populations in lakes located in both temperate regions and in the tropics can support an intensified cropping. In fact, a surprisingly high number of lakes are underfished, particularly in sparsely populated regions.

On the other hand, a great number of fish-producing lakes and rivers have been destroyed. A return to unpolluted and freely flowing waters, combined with sound management of intact rivers and lakes, would give good long-term results in more than just fish productivity. In a hungry world, where the supply of proteins is inadequate, it would certainly be worthwhile for many countries to restore aquatic habitats to where they again become productive, enabling fish to reproduce naturally.

In developed countries, the recreational value of inland waters is immense. A United States federal program to restore, perpetuate, and manage fish for sports angling purposes is financed from a 10 percent excise tax on fishing equipment. Since its beginning in 1952, $75.5 million have been apportioned for benefits that include 222 fishing lakes with a total surface of 28,000 acres. More than 60,000 acres of land provide access to these and other waters at 950 developed sites, opening 2,250 miles of streams and 2.9 million acres of bays and reservoirs to anglers. This restoration work in the United States is impressive, but the main fight is still against pollution. For many years ahead it will be the gravest conservation problem in North America and Europe. If pollution of rivers and lakes could be checked, their waters would be self-restoring and would again support self-reproducing fish and other life forms.

Aquatic Reptiles

Many reptiles living in freshwater habitats constitute a valuable natural resource that should be carefully managed. Animals like crocodiles, alligators, and the South American river turtle (*Podocnemis expansa*) are ecologically and commercially important. All these groups have been depleted by overexploitation, and local extermination.

For example, the Nile crocodile (*Crocodilus niloticus*) of Africa, in addition to its direct economic value to man, is a useful species in controlling certain predatory fish living in lakes and rivers. The predation of this crocodile maintains a balance within the aquatic community, which favors food fish important to man. Where the Nile crocodile has been exterminated, commercially important fish have declined, and unwanted species have multiplied. Thus destruction of the crocodiles does not benefit fishery interests, but on the contrary has harmful effects. It is tragic that the Nile crocodile has been exterminated in many African waters, as have a number of other crocodile species elsewhere in the world. Recently, all three species of African crocodiles have become totally protected in the Congo.

Campaigns to exterminate crocodiles undertaken in various countries constitute unwise actions working against human interests. Crocodile skins have great value, and the meat provides proteins. Profits from crocodiles are high, as illegal hunters, who slaughter large numbers of crocodiles indiscriminately, know well. The skins are smuggled out of the country, where the demand is great. Trade in crocodile and alligator skins has become an industry in Africa, South America, and elsewhere. This is a challenging conservation problem that calls urgently for a solution in the interest of the countries where crocodiles and alligators live.

The South American river turtle is a primary food resource that has been utilized for more than three hundred years. It was the basis of the economy for the human population along the rivers of the Amazon and Orinoco water systems. Exploitation of the turtle has been so reckless that it has been exterminated in many areas and has become very rare in others. It needs strong protection, so that the turtle population can be built up to its former numbers and again cropped and utilized as an important natural resource.

Aquatic Birds

Bird populations are often numerous in freshwater habitats such as eutrophic lakes, marshes, and swamps. This is chiefly explained by the abundance of food and nest sites. Large numbers of niches are often available, enabling various species to avoid interspecific competition. Some species feed on plants, others on animals utilizing shallow or deep waters, the littoral, pelagic, and bottom zones, and so forth.

The productivity of marshes and ponds for ducks has been investigated in a few areas. In Idaho, nine species of ducks produced nine young per acre and the Canada goose about 0.04 young per acre. In Utah, nine species of ducks produced over six young per acre. These figures have to be multiplied many times to give an indication of the productivity of the whole bird population in a marsh, which comprises not only ducks but also egrets, herons, bitterns, grebes, loons, cormorants, gulls, terns, coots, waders, and others. These other species, however, are generally not used either for food or for recreational shooting.

Many of these birds do not breed in or beside the lake or the marsh that contributes to their production. Freshwater habitats often provide a food supply for a number of birds that occupy the surrounding regions. Herons, loons, and other birds may fly back and forth several times a day between the breeding and fishing sites, which may be located many miles apart.

In addition, for migrating water birds and particularly wildfowl (including swans, geese, ducks, coots, and waders that fly thousands of miles every spring and autumn), lakes, marshes, and swamps must provide a network of available habitats as breeding grounds and for feeding, resting, moulting, and wintering purposes. Many of these activities can be carried out in the same habitat, but usually they are seasonally different, due to climatic and other ecological changes in the yearly cycle. Moulting is a very critical period for swans, geese, and ducks, for they are not then able to fly. Many species when moulting congregate in lakes where they are undisturbed and find food. Such lakes often have been used annually by ducks for a long time, and their location has become a part of the migration pattern of many species. This illustrates the importance of large numbers of freshwater habitats of various types existing within the geographical range of wildfowl species. The breeding areas are of as much importance to the birds as their winter quarters. They must have both to live, and moulting grounds, as well as feeding and resting areas along the flyways, are almost as important. Hence, effective wildfowl management can hardly be carried out other than internationally. Canada, the United States, and Mexico have shown the way by successfully coordinating hunting seasons and other wildfowl legislation. The European countries, through the efforts of the International Wildfowl Research Bureau, are working on a European-Asian-African Convention aimed at conserving wetlands and coordinating wildfowl legislation. This Convention was adopted in 1971.

Reserves and refuges are extremely important for wildfowl conservation. Most species respond quickly to such measures by increasing their numbers. They adapt even to minor environmental help, are flexible in requirements and habits, and build new traditions readily. In the United States, for example, many federal and state reserves and refuges now have breeding populations of Canada geese (*Branta canadensis*) where this species had not previously existed, at least for scores of years.

While modern civilization has changed and destroyed many habitats for aquatic birds, it has also created new ones. Reservoirs, canals, mining subsidences, sewage farms, gravel pits, park lakes, and other artificial habitats often attract ducks and other wildfowl for breeding, feeding, and wintering. Thanks to their ecological flexibility these birds learn to utilize all sorts of waters. Wildfowl are an easily managed natural resource of great economic, recreational, and aesthetic value that man can readily afford to have around him.

The most productive wildfowl areas of North America are the prairie potholes and marshes, the northern watersheds and deltas, and the northern forests and tundra. The prairie pothole region makes up only 10 percent of the total waterfowl breeding area of the continent, but it produces 50 percent of the duck crop in an average year and more than that in bumper years. Unfortunately, the Canadian and Alaskan deltas and wetland flats are threatened by dam projects and oil fields. Three-fourths of the Saskatchewan River delta will be defunct or seriously damaged by the mid-seventies; the Yukon is endangered, and so are the Mackenzie and Old Crow flats. If exploitation of the Far North continues at this accelerated pace, waterfowl habitats and populations will shrink considerably. As noted, the proposed Rampart Dam project in Alaska, which would have destroyed enormous areas of productive wetlands with catastrophic consequences for wildfowl, has been wisely abandoned.

The Great Plains wetlands region in Canada is also an extremely important breeding area. Though it represents only about 10 percent of North America's total duck breeding country, 70 to 80 percent of mallards (*Anas platyrhynchos*) and pintails (*A. acuta*) are produced there. The same region is the only significant breeding area for ruddy duck (*Oxyura jamaicensis*), canvasback (*Aythya valisineria*), and redhead (*A. americana*). Hence, Canada and Alaska produce wildfowl populations that annually spread over almost the whole continent. The responsibility of Canada and Alaska to conserve productive wildfowl habitats is great, but these are not the only custodians. Other countries and states of North America must contribute by maintaining their wetlands to receive the migrating and wintering flocks, which are a continental property.

Wildfowl surveys in Mexico have revealed that 10 to 20 percent of the continental duck population winters or passes through that country. A comparable condition holds true in Eurasia, but the situation there is politically much more difficult, due to the great number of countries. In the Old World the U.S.S.R. holds the key position in wildfowl conservation, because that country produces more geese and ducks than all the other Eurasian countries together. In 1965, about thirty million geese, ducks, and coots were shot within the U.S.S.R. The colossal figure dramatizes the importance of wildfowl as a protein resource. This group of birds should be considered in quite a different category from recreational game. They are much more valuable than many people realize. To repopulate with geese and ducks all the lakes, marshes, and bogs that at present do not support optimal numbers of these birds would be economically worthwhile.

Countries like Norway, Sweden, and Finland, for example, have a tremendous ecological potential for producing ducks, geese, swans, and waders. They have many suitable coasts, vast archipelagos, and innumerable lakes and marshes of different types. The population densities of many wildfowl species breeding in Scandinavia and Finland do not seem to be in accord with the great abundance of suitable habitats. Probably several factors account for this situation, some of which we do not yet know. Some we understand fairly well. We know, for example, that hunting pressure is very high on many ducks and probably also on geese. Shooting is by far the most important mortality factor. In the case of Swedish mallards (*Anas platyrhynchos*), for example, the proportion of total mortality caused by shooting is estimated at 71 percent—76 percent for inexperienced first-year juveniles, 64 percent for adult birds—and the actual figure may well be 80 percent. It is estimated that about 30 percent of the mallards shot are wounded and not recovered. Even a mortality percentage of 70 percent caused by hunting is extremely high. Apparently this pressure on the mallard populations, as well as on other species of ducks and geese, prevents their reaching a population level that corresponds to the carrying capacity of Fennoscandian waterfowl habitats.

If sound principles of game conservation can be applied in Fennoscandia, it will mean that man, taking accurate, practical measures based on biological facts, can increase the yield of wildfowl and game by rational use of the game-producing potential. Soon such measures will be the only way to safeguard future hunters' sport. Growing human populations, especially in the cities, increasing leisure, and greater mobility indicate that before long we shall have to face great problems concerning where and how people will find recreation. The number of hunters and anglers is increasing like an avalanche. Simultaneously, the general interest in animals and nature has increased enormously. So has scientific activity in the field. This is a sound development, but it is conservationists who must closely follow what is happening and lead in the right direction.

Many of the thousands of lakes and marshes in Norway, Sweden, and Finland can produce ducks and geese in much larger numbers than they do today. The qualifications—food, shelter, and breeding habitats—exist in many areas, but few or no birds breed there. We know, for example, that it is relatively easy to restore populations of *Anatidae*. Several instances of this are to be found in Fennoscandia, both in marine and freshwater species. Knowing the unnaturally high mortality caused by man in duck and goose populations, it is obvious that the simplest way to allow many of the Fennoscandian lakes and marshes to produce optimal populations of certain wildfowl species would be to regulate hunting pressure, perhaps by a totally closed season lasting several years. Such a closed season would make it possible to renew a capital that would then yield a good return in the form of game birds. The eider (*Somateria mollissima*), a coast bird, is a good example of how local and partial protection over a period of years has increased the population considerably.

Even though the problem is international, that is no reason to sit passively waiting for initiatives from other countres. Fennoscandia is part of the production area, and measures to increase the populations of ducks there will benefit these countries as well as others. It cannot be much of a sacrifice to abstain from autumn duck-shooting for a three to five year period, when the result would probably be a great increase in Fennoscandia's stock of ducks. The increased capital would then have to be managed to keep the yield at a permanently high level. Although the success of these measures cannot be guaranteed, the evidence is so strong that the outcome would be as indicated that it seems foolhardy not to try the experiment.

It is often claimed that geese and some ducks are antagonistic to agriculture, because a few species feed seasonally to some extent on farmlands. This problem has been investigated. In most cases, geese do not damage crops. On the contrary, they are beneficial to farmlands, contributing to a rapid turnover of organic matter that improves and conserves the soil. Wild-goose manure is returned to farmlands in small quantities of plant nutrients. To a greater extent the geese contribute accumulations of fertilizing compounds in freshwater habitats.

The recreational value of waterfowl has already been noted. Recreation benefits the regional economy as well as the individuals who enjoy it. This is clearly shown by experience in the United States, where three hundred national wildlife refuges encompass 28.5 million acres. These refuges are primarily for wildfowl. Ducks and geese congregate there in large numbers. Human visitors who come to observe the birds have been increasing even more rapidly. At the present writing fourteen million bird watchers, hunters, photographers, anglers, and students visit these refuges annually. The number will soon swell to more than twenty million. The same trend holds in Europe, where wildfowl reserves attract visitors by the thousands. Wildlife sanctuaries are more and more becoming human sanctuaries as well.

Aquatic Mammals

Many mammals living in such freshwater habitats as marshes, lakes, bogs, and rivers produce furs of economic value. These include muskrats, nutrias, beavers, minks, fishers, and otters. Capybaras provide excellent meat. So do grass-eating freshwater manatees, like their marine relatives. One of the world's largest land mammals, the hippopotamus in Africa, lives partly in the water. It spends almost the whole day in lakes, pools, and rivers but feeds on land by night, and contributes immensely to the productivity and fishery yield of a lake. It is one of Africa's most valuable food resources, and if wisely managed, can play an important future role in the continental economy. This is likewise true of many semiaquatic antelopes.

All these mammals fit in well along with water birds. A great number of species use the hippopotamus as a perch. Muskrats and ducks are inseparable on most North American wildfowl ranges. The muskrat's feeding and house building generally improves marshes for wildfowl. The nutria helps to maintain a balance between open water and marsh vegetation, to the benefit of wildfowl. Beaver activities contribute to water conservation and are of significant value to wildfowl.

In general, there is no reason to eliminate mammals from freshwater habitats. They comprise a beneficent element in the aquatic biocommunity, just as do fish, amphibians, reptiles, and birds. Many of them can be harvested and will yield sustained profits.

CHAPTER FIVE

The Soil

The soil is a renewable natural resource of an importance equal to water. Soil is a necessary prerequisite for terrestrial life. It is the most vital component of land and is the basis of an organic cycle that makes life possible. Without life, on the other hand, no productive soil would exist. The decomposition of organic matter is a condition of existence for soil productivity. Hence the richness of the soil comes from life both within and on it. Most soil is fertile, which is to say it is teeming with life. In fact, the biomass in productive soils is usually higher than the one existing on it, whatever the habitat may be. For example, the biomass of subsoil animals often exceeds that of the cattle grazing on the surface soil above. Collectively, the biomass of soil organisms constitutes more than half of all living terrestrial matter.

Erosion is the gravest danger to which the soil can be exposed. But erosion is also a natural process of great value. Without the geologically slow wearing down of mountains and highlands, there would be no productive valleys, plains, and deltas. Nutrient materials and soil are constantly carried by rivers and streams that flood and so fertilize lands, and that nourish fish populations in the lakes and seas into which they flow. In natural habitats erosion is a positive agent, a part of the landscape dynamics. Only where man has made an area bare of vegetation or has thinned out the bush cover or has cut down the forest, does erosion by water, wind, and frost remove soil and accelerate its effects so that it becomes harmful. Under those circumstances erosion can turn previously fertile lands into deserts. Tragically, this is precisely what happens on an increasing scale all over the world every year.

It is not a new phenomenon. For more than three thousand years man has been responsible for this kind of destruction, and he has carried it to all continents.

Recently, he has also begun to destroy the soils by spreading toxic chemicals. Many of these pesticides are persistent. The toxic residues accumulate in the earth and kill or damage the organisms living in it—and eventually kill the soil itself.

Soil Formation

Soil formation is a complex process of interaction among components of the atmosphere, lithosphere, and biosphere. It takes many centuries to create soils that can support constant vegetation, which in its turn is the basis for animal life. The litter breakdown nourishes and forms the soil, but the variations in the breakdown processes are endless, involving factors of air, light (i.e., sun energy), water, moisture, temperature, vegetation, and fauna composition. The rate of breakdown and disappearance of organic matter is greatly influenced by the number of bacteria on and in the soil. Also, primary rocks can be decomposed, but the mechanics of this slow process is not yet fully known. Water containing acids contributes to the decomposition of igneous rocks. Elements such as calcium, carbon, nitrogen, phosphorus, sulphur, and silicon are involved in long and complex geological soil cycles, within which they undergo shorter and repeated biological cycles (in annual, seasonal, or daily twenty-four-hour rhythms), passing in and out of plant and animal bodies. Soil cycles of nutrient salts are primarily biochemical.

The dynamic soil formation and maintenance system is so complex that it is difficult to summarize adequately in a few paragraphs. Although the principal factors—e.g., parent rock, topography, climate, vegetation growing on the soil, and animals modifying and using the vegetation—directly influence the soil-forming processes, the rate of soil formation differs considerably in various habitats. Within the same climate, the soil development in a rain forest is very rapid, while in secondary vegetation or on open cultivated land it is slower. To form a soil cover a foot thick may take from three hundred to one thousand years, and formation of a cover of 6½ feet of fertile topsoil may require anywhere from two thousand to seven thousand years.

The product of the soil formation process combines inorganic, organic, and living elements. As in every habitat when an optimum is achieved. the soil continues to be dynamic rather than static, even though its general physical composition appears to have become stabilized. The equilibrium is maintained by continuing dynamic processes.

Soil Groups

Soils are classified in various ways. It is unnecessary to discuss this in detail in this book, but the reader of the chapters that follow will find it useful to understand that habitats often reflect different types of soil. Residual and transported soil types form a mosaic pattern over the continents. The transported category comprises sand dunes, wind-blown loess, glacial and marine deposits, alluvial sediments, talus, and volcanic soils.

Erosion

The soil is the basis on which the whole ecology and well-being of lands and landscapes balance. The complexity of soils and of the processes that form them, as well as of the many interrelationships of soil with the physical and biological environment, makes them and their structure very vulnerable both to direct mistreatment and to misuse of the vegetation and the fauna.

As noted, accelerated erosion throughout the world is almost always caused by man's misuse of land. Altering or destroying natural vegetation to accommodate what are considered human needs is often the first negative step to unnatural erosion. Continuous misuse by overexploitation of soils through mechanical treatment, overcultivation, overgrazing by livestock, excessive cutting and burning of forests, and so on, accelerates the deterioration and leads to dreadful ravaging of the land.

Erosion of soils that have been laid bare of vegetation operates in several ways. Winds blow away fertile soil layers. During dry seasons these are carried in the air in great quantities, often in the form of dust storms. During rainy seasons the water washes away the surface soil from slopes or floods the flat plains. The runoff digs waterways and gradually cuts deep gullies, which increase the erosive processes. Data from Rhodesia show that where bare soil is exposed to the rain, the annual soil loss varies from 25 to 225 tons a year from each acre on slopes of 3 percent to 6½ percent. Under the same conditions soil loss from a grass-covered area did not exceed half a ton per acre.

During the past century water and wind erosion have destroyed an estimated 5,000 million acres of soil in various parts of the world. That is the equivalent of over a quarter of the earth's farmlands.

Particularly exposed to erosion are croplands, overgrazed pastures, unprotected forest soils, strip mines, roads, road banks, and all types of bulldozed areas. In fact, agricultural development increases land erosion rates up to ten times over what they are for natural cover, while construction may increase the rate a hundredfold. Sediments carried by man-made erosion represent the greatest volume of wastes entering surface waters. In the United States the volume of suspended solids reaching the waters is at least seven hundred times greater than the total sewage discharge loadings. A 1969 report from temperate United States estimated the average sediment yield during a rainstorm at highway construction sites at about ten times that for cultivated land, two hundred times that for grass areas, and two thousand times that for forest areas.

Plant cover eliminates the erosive effect of the raindrops hitting the bare ground and greatly reduces runoff and water loss. Erosion by frost occurs mainly in exposed moist soils. Since ice takes up more space than the water from which it is formed, soil particles are forced apart as they freeze. The loosening of the soil initiates its rapid removal by wind and water. Measurements in the Snowy Mountains of Australia have shown that such frost actions can remove as much soil in a single winter as would require at least six hundred years to form. Here again, frost erosion cannot remove soil unless the activities of man first destroy the vegetation that protects it.

In well-balanced ecosystems, where forest and other kinds of vegetation protect mountain slopes and habitats beneath the trees, negative erosion does not occur, and the hydrography of the country functions in a positive manner that contributes to the fertility of the soil. Man should take advantage of such optimal situations, which are the basis of sustained yields whatever the land produces.

The conservation of soil resources is among the most important tasks confronting governments all over the world. On all continents except Antarctica erosion is a major hazard to the future of mankind, but few countries seem to be aware of how serious the situation is. If they are, they do not take effective action. While tragic, this is perhaps not so remarkable, for the catastrophic situations are the result of continuous misuse of land for thousands of years. The case is man alone, though his livestock, primarily the goat, have helped. The climate cannot be blamed. For example, in the first Israeli-Arabian conflict some areas of the Middle East where the soil was seriously eroded and desertlike were evacuated. After some time, with the pastoralists and the goats gone, plants that had not grown there in the memory of man began to reconquer the wounded soil.

Flying over Mediterranean countries leaves a memory of a mountainous desert in which erosion has made ravines like great gashes in the reddish-brown earth. Here is a landscape literally murdered, where rivers on their way down from the mountains have torn up the defenseless earth, and wind has blown away the soil. Now the river beds are dry and bare, for even during the sudden floods of spring and autumn the waters do not bring them to life but only eat still further toward sterile bedrock. Not even lichens are given an opportunity to conceal the jagged rocks.

A warning of what may happen on the European side of the Mediterranean is provided by neighboring Africa, where the Sahara Desert reigns. Indeed, the desert is now spreading over larger and larger areas of Mediterranean Europe. In the mountains of Italy, both the Apennines and the Alps, 80 percent of the pastures are damaged by serious erosion, primarily caused by overgrazing. Here, too, having lost their protective vegetation cover, the slopes lie naked. Even worse is the situation in the Sierras and high plateaus of Spain and in the mountains of Greece. Two to three thousand years ago, many areas of the Middle East were fertile lands supporting flourishing civilizations, mighty kingdoms, and large populations. Today the same areas are pure deserts, where only the ruins of great cities provide reminders of former glory. In northern Syria alone, man-made deserts comprising more than a million acres have buried more than a hundred cities. Nothing now but stone skeletons, these cities died because the land around them was killed by the inhabitants. Babylon met the same fate. So did the cities of ancient Mesopotamia and many other countries with a glorious past.

No natural deserts exist in India, but in recent time desertlike tracts have come into being—a consequence of human misuse of vegetation and soil through overgrazing by cattle. For the same reason, large areas of African semideserts and savannas that formerly supported a wealth of wild animals have turned into sterile lands. There are many indications that the spread of the North African deserts is primarily due to cutting of trees and overgrazing by cattle and goats. During Phoenician, Roman, French, and Italian civilizations in North Africa, when the governments were stable and progressive, vegetation slowly reconquered the desert. But every time the Arabs returned to power, nomadic tribes and their livestock invaded the fertile areas, which soon collapsed due to misuse, and deserts again took over. The same phenomenon is to be seen in the southern Sahara, at the edge of the savannas. Likewise, in many parts of East Africa, West Africa, South Africa, and Madagascar, the deserts are spreading in previously fertile country with alarming speed. As soon as overgrazing by cattle, goats, and sheep occurs, or farmers practice unwise agricultural methods, productive savannas with comparatively rich vegetation and wild animal life rapidly disappear and give way to deserts. Dunes that have been kept in place by the vegetation for centuries revert to moving sand waves, killing everything in their path. River sources dry up, and so does the land along the dry river beds. The rains, no longer stored by the vegetation, sink deep down into the earth or run off, carrying fertile soils down to the sea.

Proportionately no continent has suffered so much from erosion in so short a time as Australia, where one mistake has led to another during the two hundred years of European colonization. In North America the erosion process has also been very rapid, due to deforestation and particularly to overgrazing and mechanical cultivation of the central and western plains.

Again and again the story is the same. Study of the prime causes of erosion damage shows that practically all over the world it has been preceded by overgrazing and overtrampling by livestock or by unwise methods of cultivation—kinds of land use that tended to remove and destroy the vegetation cover. The results are that the whole ecology of the ecosystem has been disrupted.

One of the most terrible consequences of erosion is its influence on the local climate and hydrography: aridity and lowering water tables often follow in its steps. Water is no longer stored in upper soil layers and becomes unaccessible to plants and animals. Dry seasons, which previously were a normal thing, tend to develop into drought catastrophes, embracing a much wider region than the eroded area.

The ruthless exploitation of the soil clearly cannot be allowed to continue. Humanity has long since reached a point where it was consuming more of the earth's soil capital than nature could regenerate or man himself could help to restore. This situation has a bearing on the existing food shortage in many parts of the world. Effective soil conservation is necessary. It must look ahead, employing an ecological perspective not only in order to preserve and utilize wisely but also for building up and increasing productive soil resources. This cannot be done without conserving vegetation.

The United States bases a considerable part of its economy on agriculture and is considered to be one of the richest agrarian countries of the world. By the end of the 1940's most of the cultivated soils of the United States had lost more than half of their long accumulation of organic matter. Even worse is the loss of the soil itself. More than 100 million acres of the fertile top layer of cultivated soils have been destroyed by erosion in the United States. That is more than one-sixth of the country's agricultural land. Most of the American soils are ruined when six of the topsoil's approximate ten inches have been lost. This is happening over increasingly large areas. In many locations farmers are tilling the much less productive subsoil. This deteriorative process has taken but a few hundred years in the United States—less time than in any other area of the world. During the dust bowl disasters in Colorado, Texas, Kansas, and Oklahoma in the 1930's, the loss of topsoil ranged between two and twelve inches. The windblown sand-dust damaged millions of acres of farms, and drifting dunes buried vast areas. Land and people were ruined. It was a hard lesson, entirely caused by human misuse of the soil. The U. S. Soil Conservation Service presently estimates that at least 14 million acres of the Great Plains currently under cultivation should be returned to grasslands. Whether the plains have ever produced as many proteins as when they were grazed by bisons and other wild ungulates is a moot question. In any event, one thing is sure: during that not distant time past, when millions of bisons roamed the North American plains, the soil did not deteriorate.

In Europe, too, the productive life of the cultivated plains is slowly deteriorating. The spring floods have been canalized. They no longer nourish the meadows or provide storage water to the marshes or the soil. The groundwater level is sinking.

In 1965 tremendous dust storms occurred in South Africa during a drought period. Inland dust was carried coastwards in the upper air layers and caught up by the rain there, resulting in mud showers over the Natal coast. South Africa has had a Soil Conservation Act for more than twenty-five years, but the existence of this tool for controlling erosion has not kept the soil from deteriorating, the desert from encroaching, vast quantities of productive topsoil from being carried to the sea year after year.

Flying over Africa and Asia is another shocking experience. Apart from the deserts, bare soil exists over enormous areas. The annual loss of topsoil must reach astronomic figures. Using data from Rhodesia as a guide, a loss of twenty tons of soil per year from each acre of bare ground is a conservative estimate. At that rate, it will take only fifty years to lose six inches of the productive topsoil and less than eighty years to lose the whole fertile top layer. These calculations represent minimum figures. In reality it is likely that the loss is much more rapid. Similar examples of rapid soil loss are also to be found on other continents.

Does mankind realize what this continuous soil destruction means? One of the world's most precious natural capital resources is being tapped continuously without replacement. In addition to the direct loss to agriculture, dangerous imbalances of productive natural habitats and ecosystems are resulting. The condition is a menace to all life, including man. It is manifest not only in the periodically drought-stricken areas of Africa, Asia, Australia, and the Americas but everywhere throughout the year, because the general productivity is lowered, the soil cycles are upset, and the soil formation does not function. The situation constitutes a death threat. Only adequate, rapid conservation measures employing engineering and biological methods can give us hope of repairing the damage—and nature a chance to heal the wounded soils. They will be costly. They will take time. But they are necessary for human survival.

Many means for halting erosion and improving soils are at hand. Most important and urgent is restoration of the vegetation cover in order to stop the erosive actions of wind and water. Any kind of vegetation can do the job—even species that are usually considered to be pestiferous. It is better to cooperate with such plants than to surrender to erosion. The next step is to let grass and forests grow again on lands that have become too arid and poor to support cultivated plant life. Reforestation will also help to restore the local climate. Another means is to utilize the water resources properly. Mechanical and engineering measures such as terracing, water reservoirs, and the like may be necessary additional aids to restoring the soil, but they cannot alone achieve positive results. They must be preceded or accompanied by basic biological measures.

The ultimate goal of soil restoration and conservation is to put the biological processes of soil formation and soil life cycles back to work as they functioned before the vegetation was destroyed. This is a long-term task. It may well take five hundred to a thousand years or more to restore forests that once grew on mountain slopes. These forests, as well as lowland forests and grasslands, over the course of millennia had themselves created the soil humus on which they based their existence. And before the trees colonized the land, other types of vegetation—also during thousands of years—had prepared the ground for successive plant communities to evolve, building up to and producing a maximum conversion rate or energy flow in which the soil was an imperative force. Then man destroyed the whole climax structure—a habitat with relatively stabilized environmental conditions—ruining not only the landscape but also his own economy. This is what has happened and is happening increasingly over vast areas of the world.

In many countries, particularly in the tropics and subtropics, destructive land use and increase of land-destroying livestock, as well as human population growth, hamper the effects of antierosion measures to such a degree that the soil erosion accelerates despite the combat against it. The prospects of the fight for survival in such areas are not bright.

Soil Pollution Through Solid Wastes

Various forms of pollution affecting the soil are discussed in the chapters on the air, the sea, fresh water, vegetation, and animals. There is another form of man-made pollution which has an impact on the landscape as a whole and therefore is difficult to place in any particular chapter. It is the dumping of solid wastes, which perhaps is more directly linked to the soil than to other renewable natural resources. Such refuse in industrial countries is found almost everywhere as a normal feature of the landscape, and in the developing countries the same deplorable situation is beginning to take shape.

The effects of solid wastes accumulating on the soil are visible and, therefore, easy to realize, but they are only a part of the quantity, because large amounts are dumped in the oceans, in lakes and in rivers, without thought of contamination of ground water tables.

The most significant data on the magnitude of the solid wastes in an industrial country are available in the United States and are found in the recently published (1970) First Annual Report to the U. S. Congress of the Council of Environmental Quality, from which the following information is quoted.

> The growing technology and affluence of American society have laid a heavy burden on solid waste facilities. Refuse collected in urban areas of the Nation has increased from 2.75 pounds per person per day in 1920 to 5 pounds in 1970. It is expected to reach 8 pounds by 1980. This spiraling volume of solid waste has a changing character. The trend toward packaged goods in disposable containers has put more paper, plastics, glass, and metals instead of organic matter into the refuse. And technology of solid waste collection and disposal has not kept pace with this change.
>
> The total solid wastes produced in the United States in 1969 reached 4.3 billion tons as shown in the following table:

Residential, Commercial and Institutional wastes	250 Million tons
Collected	(190)
Uncollected	(60)
Industrial wastes	110
Mineral wastes	1,700
Agricultural wastes	2,280
	4,340

> Most of it originated from agriculture and livestock. Other large amounts arose from mining and industrial processes. A little under 6 percent, or 250 million tons, was classified as residential, commercial, and institutional solid wastes. And only three-fourths of this was collected.
>
> Although wastes from homes, businesses, and institutions make up a small part of the total load of solid waste produced, they are the most offensive and the most dangerous to health when they accumulate near where people live. Agricultural and mineral wastes, although much greater in volume, are generally spread more widely over the land. They are more isolated from population concentrations and may not require special collection and disposal. Nevertheless, as more is learned about the effects of agricultural and mineral wastes on the quality of air, water, and esthetics, steps to curb their production and facilitate disposal seem likely.
>
> The largest single source of solid wastes in this country is agriculture. It accounts for over half the total. The more than 2 billion tons of agricultural wastes produced each year includes animal and slaughterhouse wastes, useless residues from crop harvesting, vineyard and orchard prunings, and greenhouse wastes.
>
> Herds of cattle and other animals, once left to graze over large open meadows, are now often confined to feedlots where they fatten more rapidly for market. On these feedlots, they generate

enormous and concentrated quantities of manure that cannot readily and safely be assimilated by the soil. Manure permeates the earth and invades waterbodies, contributing to fish kills, eutrophied lakes, off-flavored drinking waters, and contaminated aquifers. Feedlots intensify odors, dusts, and the wholesale production of flies and other noxious insects. Animal waste disposal is a growing problem because the demand for animal manure as a soil conditioner is declining. Easier handling, among other advantages, favors chemical fertilizers.

The solid waste collected annually includes 30 million tons of paper and paper products; 4 million tons of plastics; 100 million tires; 30 billion bottles; 60 billion cans; millions of tons of demolition debris, grass and tree trimmings, food wastes, and sewage sludge; and millions of discarded automobiles and major appliances.

A number of specific components in solid wastes present particular problems and require special mention. Abandoned autos are one of the most conspicuous solid waste disposal problems. On the average 9 million autos are retired from service every year. Although statistics on the annual number of abandoned vehicles are subject to dispute, it is thought that approximately 15 percent are abandoned on city streets, in back alleys, along rural roads, and in vacant lots throughout the Nation. Most autos are abandoned because they are no longer serviceable and have little or no parts value to auto wreckers. The total number of abandoned cars in the country is even harder to ascertain, but has been estimated between 2.5 and 4.5 million.

There might also be some useful environmental effects of the dumping of solid waste. Some animals, particularly birds, utilize this "solid waste habitat" for nest building, and when skin diving in Florida and Australia I have several times observed how fish take ecological advantage by adopting dumped rubber tires or other man-made objects as an important part of their territories. In fact, the Fish and Wildlife Service and the Bureau of Solid Waste Management are investigating the use of old tires as reefs and fish havens along the Atlantic coast of the United States. The ocean bottom is sandy and relatively flat for great distances. And artificial reefs constructed of tires may promote an increase in desirable species since many game fish require relief features such as reefs for protection and spawning grounds. If this concept proves practical, very large numbers of old tires could be turned into an important ecological side benefit. In eastern Canada new lobster colonies have been established by dumping rocks on the sandy sea floor to create artificial reefs.

CHAPTER SIX

The Vegetation

An area that has lost its vegetation, its plant and forest cover, is a ruin, a dead piece of land. The plant cover protects and partly creates the soil; the forests accumulate and distribute water, and modify the local climate. Without these an area cannot function as a living landscape. In addition, the vegetation is the sole means of converting solar energy, minerals, and moisture into forms sustaining animal life. In other words, the vegetation determines in relationship with sunlight, water, and soil the basic patterns of the natural environment as well as of what is today called the human environment. The preceding chapters on water and soil have tried to show how interrelated and complex a living landscape is. All components of it work together like cells and organs in a living body.

The vegetation of a region reflects historic and ecologic factors. It is a manifestation of past ages, present conditions, and environmental space-time interactions. Climatic, edaphic (soils), and biotic (plants and animals, including man) factors are involved in the contemporary interrelationships on which the vegetation and plant distribution depend. Plants are the basis for animal life and, hence, also for man's existence. He does not always seem fully aware of this fact.

Vegetation maintains life on the earth in other ways than nourishing soils and providing energy intake for animals. The green plants produce oxygen by photosynthesis. It is believed that before life evolved on earth there was no oxygen in the atmosphere. The plants filled the atmosphere with their gas. At present the vegetation cover maintains the oxygen balance. How long will it continue to do so? In a situation where man rapidly destroys the earth's vegetation cover at the same time that he pollutes the air and the sea, and the size of the human population accelerates far beyond what the environment can tolerate, the question must be asked: where and when is the critical limit at which nature will strike back? Man cannot forever escape natural population barriers. Will a change in oxygen production prove to be a decisive barrier?

The tide of man-made destruction on earth has been running for thousands of years, but at an accelerated pace during the two last centuries. It is still spreading. The easiest way to confirm man's failure to cooperate with global natural resources is to measure the decline and disappearance of the vegetation. The plant cover is a reality to everybody, easy to observe and easy to destroy. All over the world forests and grasslands are retreating. Practices such as burning, shifting cultivation, overgrazing by livestock, and ruthless timber-cutting have been highly destructive to the vegetation. Agriculture has spread over large areas where forests, woodlands, and grasslands have unwisely been cleared without any long-term benefit at all. Instead the destruction of vegetation has led to erosion, flooding, drought, vanishing wildlife —an impoverished environment.

The world's main vegetation communities or regions are often in an ecological sense referred to as biomes or biotic regions. They include a wide range of vegetation such as tundra, desert, chaparral, grassland, coniferous forest, temperate deciduous forest, tropical rain forest, and so forth. Where undisturbed by man, biotic communities, including animals, occupy particular areas and evolve to a climax formation, presenting a situation of potential productivity and maximum energy flow—i.e., energy conversion rate—within the local climatic, physiographic, and biotic conditions. It would be worthwhile for man to learn and ultimately interpret what these biotic climax situations indicate before he destroys the whole pyramid structure in order to replace it with some other form of development, which often turns out to be less productive.

Ecology of Plants

This book often refers to stable ecosystems and stabilized environmental conditions. The reader must bear in mind that these expressions refer to ecological stability of natural biocommunities, in contrast to the instability of man-made or man-influenced habitats. In reality ecosystems are always dynamic, responding to climatic and topographical changes as well as to changes in distribution, frequency, and the genetic potentials of the plant and animal species that form the structure of the ecosystem. The evolution of a natural ecosystem is a slow process. In fact, ecosystems composed of climax habitats change so slowly that it may take hundreds or thousands of years for significant changes to occur. That is the meaning of stable ecosystem and community stability. As against this normal ecological stability, man's technological advance is causing environmental changes at a furious pace, bringing on an alarmingly destructive ecological instability. Paradoxically, monocultures belong to such unstable environments, due to their very lack of diversity, while habitats with high species diversity demonstrate community stability.

The schematic functioning of ecosystems and biotic communities has already been touched upon. Within the energy flow of an ecosystem, the main role of the vegetation is to capture and store the sunlight energy and to transform it for use by animals, decay bacteria, fungi, and other organisms during various stages of their life processes. The energy relationships within an ecosystem's conversion cycle are thus based upon the green plants, which transmit sunlight energy indirectly in the form of food for herbivores. These in their turn pass on the energy in the form of meat to predators, carrion eaters, and man. All these organisms, plants as well as animals, live or dead, return to the soil, and the cycle is back to soil formation and soil mineral cycles.

This story of energy flow, growth and death, goes on continuously. Hence, the natural vegetation plays a vital role in maintaining a flourishing ecosystem, where there is a balance between losses and replacement. But this is seldom the case where man has transformed grassland or forest communities, which originally consisted of hundreds or thousands of species, into uniform monocultures of a single species. Moreover, the situation is worsened when the single species is cropped and taken away from the original site to be consumed elsewhere. And not even there does it return to the soil. Agricultural and forestry practices have ruined many ecosystems, countries, and human populations before man understood what he had done to bring on the destruction. Such ruin still continues on a large scale.

It is estimated that the natural vegetation of the world before man began his destructive work some hundred years ago was about 42 percent forest, 24 percent grassland, and 34 percent desert. Today the same figures are approximately 33 percent forest, 27 percent grassland, and 40 percent desert. These figures are rough estimates because the exact proportions cannot yet be accurately quantified. Parts of some major deserts do carry vegetation but have been included in the desert percentage above. Woodland savannas and grassland savannas are in global scale not easy to estimate separately and have here both been included in the grassland percentage. Vegetation types like tundra, montane plant communities above the timber line, and strips of chaparral (macchia) and mangrove have been excluded from the estimates. The surface of the earth is three-fourths water. We have to safeguard the relatively small area available to terrestrial life. Half of the earth's land surface is uninhabitable for human beings because it is too cold or too dry or too high.

Rainfall and temperature are important ecological factors for the distribution of plants. In many areas, particularly in the tropics and subtropics, it is important how the rains fall as well as how much it rains per annum.

Though man has altered the face of the earth quite considerably by changing the vegetation, essentially the same biotic regions from the polar caps to the equatorial belt have existed since man appeared on the globe. Their boundaries have shifted and their aspects have been modified by climatic fluctuations and man's actions. These two agents are the prime modifiers of plant distribution, but only the latter is damaging the vegetation and virtually exterminating plant species.

Tundras

The northernmost parts of both North America and Eurasia consist of tundra and treeless heaths. The term tundra is used here in a broad sense—rather than exclusively for permafrost areas—for the treeless and barren grounds stretching as a circumpolar belt between the forests and the permanently snow-and-ice-covered polar areas. Also, high montane (alpine) heath areas south of the tundra belt can be regarded as tundra. The northern polar cap is encircled by the arctic tundras. Several Antarctic islands are tundralike, and so are the borders of the Antarctic continent. Tundra regions are relatively poor in species, and their plant communities seem to be simple in comparison with the more complex and diversified plant life of more southerly latitudes. However, a surprising variation of habitats and ecological niches also exist in tundra regions.

The present tundra areas are young formations. Large parts of them have been free of major ice burdens for as geologically short a time as seven thousand years. The terrain is actually a result of the last glacial period of marine introgressions of post-glacial lakes and of rivers that deposited sediments as the ice sheet retreated. Plants and animals then invaded the area. Although the tundra seems static, this process is still going on, but it evolves very slowly.

The characteristic tundra vegetation is lichens, moss, and herbaceous plants. In summer the tundra grows a wealth of plants that nourish a surprisingly high number of animals, to which belong such large herbivores as reindeer and musk-oxen. Lichens are staple food for the enormous reindeer herds that once roamed the North American and Eurasian tundras. Lichens are vulnerable to overgrazing, trampling, and burning. They regenerate very slowly. It may take fifty to a hundred years for *Cladonia* lichens to recover after having been grazed or burnt out. Overgrazing occurs in many tundra areas where semidomestic reindeer are kept in numbers much too high in relation to the carrying capacity of the vegetation. This has long been the case in Sweden, resulting in high mortality of reindeer. Though arctic tundra does not burn easily, there are many examples of tundra fires caused by man. During dry summer periods the lichen cover can be burnt over vast expanses, as a result of fire caused by man. Where fires have eliminated the lichen cover down to mineral soil, the recolonization will take even longer than a hundred years. Until then the reindeer cannot use the area.

The dwarf-willow (*Salix herbacea*), a tiny, creeping plant of the tundra, has a surprisingly high nutritious value due to its chemical composition. In Iceland and possibly elsewhere, it is the winter staple food for the ptarmigan (*Lagopus mutus*), but this selective grazing does not lead to the exhaustion of the willow.

Tundras have long been considered the least vulnerable of habitats. In national parks and nature reserves, walking by visitors in tundras and mountain heaths has been thought to be harmless. Unfortunately this is not so.

In what is now Rocky Mountain National Park in Colorado, prehistoric man had used a ridge as a route for six to eight thousand years. The effects of this use were confined to a trail and two camp sites at tree limit. The surrounding alpine tundra area received no domestic grazing or other use until 1932 when the National Park Service opened an auto road that followed the general route of the old trail. In subsequent years the route attracted many visitors to the ridge. The activities of the visitors have significantly altered the associations adjacent to the parking areas along the road. The most destructive activity has been walking on the tundra. The effects of foot traffic on previously undisturbed tundra were observed in 1958 when the Park Service opened a new parking area but constructed no paths adjacent to it.

Visitors wandered away from their cars in two directions, and within two weeks after the parking area was opened on May 30, distinct paths were visible. The tiny tundra plants were matted, the flower stalks broken, and attrition of the cushion plants was starting. By July 25, these paths were very distinct, and the attrition of the plants was considerable. The entire area adjacent to the parking area was matted and the plants had not bloomed. By August 25, the attrition had destroyed two-thirds to three-fourths of the surface area of the cushion plants.

Elsewhere on the ridge, it is observed that the effect of intense visitor use for thirty-eight seasons has been nearly total destruction of the original association. How soon these conditions were produced is not known. All plants are gone except the few protected by large boulders. Five inches of topsoil have been eroded, leaving a bare mineral soil surface.

The effects of twenty-five years of visitor trampling on a stand of climax tundra (*Kobresietum myosuroides*) were studied in depth. The numerous species of lichens and moss characteristic of this association were totally absent and have not appeared after seven years of protection. Based on various observations of growth rate in alpine plants and on comparison of successional stands to the *Kobresietum,* it is estimated that a minimum of five hundred years will be necessary for the ecosystem to restore itself.

How rapidly deliberate or careless destruction can hit the arctic tundra environment is reported by Tom J. Cade of Cornell University, who visited Alaska in 1963 and in 1969 returned to the same areas, which since his last visit had been prospected by oil companies. His shocking story reads as follows: Erosion ditches and tracks "occur about every five to ten miles from the Canning River almost to the Colville, running in north-south lines from the Brooks Range to the coast. This unbelievably callous destruction of the terrain by geophysical crews working for the oil companies is unprecedented in the annals of arctic explorations. . . . Much of the winter road network will soon develop the same kind of erosion, which results from the melting of permafrost after the vegetative cover has been destroyed by motorized vehicles. In short, the wilderness I knew twenty years ago in northern Alaska is already a thing of the past. Except in some of the mountain fastnesses of the Brooks Range, I doubt that one can find a 100-square-mile plot of ground east of the Colville River—including the Arctic National Wildlife Refuge—that does not show some irreparable sign of man's unrelenting activities."

Another example of the fragility of the tundra comes from Scotland. During the last few years the building of new roads and ski lifts, and a consequent increase of human traffic in all seasons, has damaged the mountain tundra soil and vegetation in and outside the Cairngorms National Nature Reserve. This is unfortunate in a touristic, recreational, and scientific area of high environmental quality. It has led to erosion which is a serious potential threat to the roads and ski lifts themselves. There is also a menace to animal populations, due to direct disturbance or indirectly to habitat change (Watson and others, 1970).

Another threat to the tundra ecosystem is pollution through sulphur dioxide (SO_2) to which lichens seem to be extremely sensitive. Recent oil-field development in arctic North America may eliminate lichens over wide areas through a combination of biological and meteorological factors and uncontrolled burning of crude oil, fuel oil, and natural gas.

These examples show the extreme vulnerability of tundra plant communities and particularly of climax growths of lichens. The complex of the arctic tundra ecosystem is to a great extent based on the lichens. When they disappear by overgrazing, overtrampling, or burning, the reindeer population will also collapse, and many other ecological repercussions will negatively affect the area's productivity, including the human economy—in these instances the economy of Indians, Eskimos, and Lapps.

Deserts and Semideserts

Most deserts are not merely sands. Many deserts produce life, and some of them can even be classified as productive. In reality, deserts are not at all so monotonous as people usually believe. Like the tundras, they create a number of diversified habitats with relatively simple structures. In fact, the world's deserts and semideserts carry some vegetation most of the time and periodically, after the rains, they produce plants in large quantities. In the Sahara, the world's largest desert, vast areas are devoid of all vegetation, but irregular rainstorms—sometimes coming only after years of drought—may suddenly be followed by rapidly growing grass and scrubs that have remained dormant. Even animals appear in near-miraculous fashion. Other parts of the Sahara have a permanent vegetation of scattered shrubs, bushes, and stunted trees. Where there are fissures, ravines, wadi beds, and the water table is not far below the soil surface, the vegetation survives or has a potential to do so, but is often exterminated by domestic sheep, goats, and camels.

Extreme deserts, particularly those with drifting sands, have virtually no vegetation. On the other hand, deserts like those in North America have a surprisingly rich vegetation of treelike bushes, shrubs, and annual grasses. Sagebrush (*Artemisia tridentata*) covers many of the North American semideserts, particularly the Great Basin region between the Rocky Mountains and the Sierra Nevada range in California. Mesquite (*Prosopis juliflora*) is a typical desert tree in the United States, Mexico, and northern South America. Growing to a height of thirty feet, locally it forms real forests. In western Mexico and northern Venezuela, a desert thorn forest is composed of small, leguminous trees. Other examples of desert vegetation may be given from Asia, Australia, and even certain islands in Oceania. About 40 percent of Australia is a natural desert.

Natural deserts are climax habitats, a biological response to low annual precipitation and high diurnal temperatures. Almost all of the Old World deserts produced large herbivorous mammals in great numbers, as well as a wide range of rodents and reptiles. The latter groups are also characteristic of life in American deserts, in which deer is the largest ungulate. Thus the comparatively poor desert vegetation is normally enough to support large herds of a number of nomadic, wild ungulates, which convert the grass into meat. These wild mammals indicate the great nutritious value of the desert vegetation, if used properly.

As most deserts are climatic phenomena, it might be assumed that the steady encroachment of deserts in the tropics and locally also in subtropical areas is a result of desiccation or other climatic factors. In general, this is not so. The expansion of deserts to fringing savannas, steppes, and arid scrublands is primarily caused by man and his livestock and is the result of burning, overgrazing, and overcultivation. The negative repercussions of this misuse of land is accentuated by the climatic dry period that prevails in our time, but climate alone cannot achieve the same negative results. Where man does not interfere, the savanna reconquers the desert, and the forest the savannas. This has been observed in Africa and in Asia.

Bush and Shrublands

Great tracts of the world's arid and semiarid regions are covered by bushes, shrubs, and small trees. Some are evergreens or partially so, but the larger part is deciduous and often thorny. There is also a thin cover of grass or at least patches of grass. It is a matter of choice whether one refers to such bush and shrublands as semideserts or as arid savannas, steppes, or prairies. But in many regions of the world, particularly in Africa, Asia, and Australia, and also in the Americas, this kind of biome is distinguished as a category in its own right. However, the dividing line between bushlands and semideserts or arid savannas is not always easy to draw. Bushlands under the strong influence of fires and other human factors become transformed into semideserts in the same way that savannas deteriorate to bushlands.

Natural bushlands have peculiar conservation problems and should be managed accordingly. They are relatively productive, which is clearly indicated by the large mammals living there. However, they are extremely vulnerable to misuse, which easily upsets the interplay of moisture gain and loss. Evaporation is pronounced in regions with low relative humidity, dry winds, hot sunshine, and high temperatures. Therefore, it is important that the existing vegetation be able to make use of the rainfall in the right way. In natural habitats it generally does, but where the vegetation has been altered or partly destroyed, the precipitation in the form of light showers may be completely evaporated before it can sink into the soil and be used by plants. Or torrential rains wash away immediately in a surface runoff. In both cases the results are gradual impoverishment by desiccation and lowered productivity.

Many data indicate that semiarid grazing lands are being destroyed at an accelerating rate. This conversion of productive areas into deserts is pronounced around the edges of the Sahara and the deserts of the Middle East. It is also marked in other areas of Africa, Pakistan, India, Mongolia, and Australia. It can even be observed in western North America.

Chaparral (Maquis, Macchia)

Areas having long, dry summers and cool, wet winters develop a characteristic vegetation of dry brush and evergreen woodlands, a response to drought conditions. Such a vegetation is called chaparral in North American English, maquis in French and macchia in Italian, and the three terms are frequently used interchangeably in the technical, ecological terminology. It occurs in California, the Mediterranean region, and in the Cape Province of South Africa. All these regions are coastal and have what is called a Mediterranean climate and a Mediterranean vegetation.

The vegetation in all three of the world's major chaparral regions has been greatly modified by man. In California, Indians, Spaniards, and Californians have frequently set fire to the chaparral slopes for hunting or to create open foraging areas for livestock. Mud and soil flows were often released on hills and slopes by the winter rains, where the fires of the preceding summer and autumn had destroyed the surface cover. On the other hand, fires may also lead to a subsequent thickening of the new chaparral growth. When the next fire sweeps such thickets, it is a virtual explosion, and the result may be disastrous not only for the chaparral but also for surrounding grasslands and forests, which are later invaded by the chaparral. Natural lightning has been a normal fire hazard in the chaparral for thousands of years, but the human fires have been so frequent and often set and timed wrongly that they have destroyed rather than maintained. Fire control in Californian chaparrals is essential. Fortunately it does exist, but fire should also be used as a management tool for land use.

An example of the devastating effects on vast areas of the chaparral country was a large brush fire which began on November 13, 1970, and raged out of control for four days in Los Angeles and San Bernardino counties. Eighty-one square miles of brush land were destroyed. The economic value of the damage to the watershed is placed at $16 million, to which must be added the values of lost property (fifty-two homes were destroyed), of recreation, and future damage from floods and mud slides.

The chaparral vegetation of California is regarded as having little or no economic value. However, it plays an important role as a hydrographic and ecological stabilizer preserving watersheds and the soil. Morever, it is an excellent and productive deer country.

In the Mediterranean regions, devastation of the vegetation has led to deserts where once there was chaparral, and to chaparral where formerly forests luxuriated. It is difficult to imagine how Mediterranean Europe, Asia, and Africa once looked. During the past three thousand years the soils have been dissipated, and although the climate has hardly changed, its effects are no longer what they were when the land was rich in forest. Today the Mediterranean countries are living on a rapidly diminishing capital, with fertile soil being washed into the sea or blown away. Unless a radical change occurs soon, all the land is doomed to exhaustion.

The Mediterranean countries were the cradle of Western civilization, and no place on earth has witnessed so many historic events in the last three millennia. The chronicles of these events reflect the natural environment of each period. They help us reconstruct the slow but dramatic changes that have turned fertile, flowering coasts, plains, and mountains into a ruined landscape. The story is a depressing record of biological and economic misuse, for which man alone is responsible. It demonstrates how profoundly the rise and fall of cultures depends on natural conditions. The Hellenic epoch and the greatness of Greece were based on natural resources and persisted as long as the forests produced timber and the equilibrium of nature was preserved. When the balance was disturbed by ruthless exploitation of the forests and overproduction of cattle and goats, the power of Greece declined.

The same destructive course followed in the Roman Empire, though it took longer—nearly a thousand years. As the soil of Italy lost its fertility and could no longer support the people of Rome and other towns, the empire reached out to include all the Mediterranean countries and many others. The Mediterranean became the *Mare Nostrum* of the Romans, and the whole of this mighty empire became the larder of Rome. This golden age, the *Pax Romana,* was the longest period of peace in the history of the world. When at last the Roman Empire fell, the productivity of the Mediterranean area had fallen off greatly, and the causes were the same as those that had led to the decline of Greece.

Man has not allowed nature to recuperate in the Mediterranean region. The misuse of natural resources has continued without interruption, and the decline of arable lands has not stopped. A vast amount of erosion has resulted, leaving the soil bare and dead.

Whether it was a change in climate or the action of man that caused the impoverishment of the Mediterranean region has long been the subject of debate. Most of the evidence supports the view that the desolation is man's work. For six or seven thousand years, the climate of the Mediterranean area has been largely constant. (Although a somewhat higher precipitation occurred during the last century before Christ, such fluctuations do not deserve the name of climatic changes.) The conclusion that may be drawn is a very important one: if no clear, long-term climatic change has taken place in the thousands of years during which the natural wealth of the Mediterranean countries has been squandered by man, the present development toward a sterile, desertlike condition can be halted. We may even dare to hope that this dead landscape can be revived, the dying soils saved, and the vanished forests restored. The climate clearly would not prevent a return to former natural conditions.

The chaparral of South Africa is confined to a rather narrow stretch of the Cape coastland. It contains marvelous floral wonders of many endemic species and is a great tourist and recreational attraction during the flowering season in spring. Regrettably, excessive fires, urban development, and a mania for introducing exotic plants and animals have greatly disturbed and changed the natural vegetation, which South Africa is scientifically and ethically responsible for preserving intact.

Grasslands (Savannas, Steppes, Prairies, Pampas)

Temperate and tropical grasslands provide pastures for grazing animals. They have been much used for growing agricultural food plants and have been the basis for early civilizations. Unchanged savannas, steppes, and prairies once supported biomasses that in the form of animal protein were the highest ever produced in terrestrial habitats. Man himself has been intimately related with grasslands since he emerged from the forests of Africa about two million years ago. Despite this long experience, he has failed completely to understand the ecology of savannas, steppes, and prairies. He has devastated few other biotic regions so thoroughly. Even today he continues to convert large areas with productive soils and vegetation into useless deserts. In the Old World the destruction has gone so far that in many regions it now appears to be irreversible. In the New World man has had less time for destructive work, so there the damage may be repairable through sound range management.

Natural grasslands evolve where rainfall provides conditions that are intermediate between those governing deserts and forests. Temperature, edaphic, and biotic factors are also involved, for example, grazing of ungulates. In subtropical and tropical regions, fire is an important factor for forming and maintaining grasslands. Natural ignition by lightning occurs but is negligible compared to man-made fires. In the Albert National Park in the Congo, Africa's oldest national park (established 1925), man-made fires have been brought under control, but they reach the reserve's periphery from outside areas. In the interior of this national park, fires caused by natural lightning have occurred every third year on the average, but this does not mean that the same area is always burnt. In Africa, man-made fires have swept the savannas for thousands of years—they have almost become a natural force. It has even been postulated that tropical savannas are a product of man-made fires. It may be partly so at present, but the fact that Africa has evolved a well-adapted, specialized, and diversified savanna fauna, tremendously rich in species, clearly indicates that savannas are an old biome that existed long before man. Therefore the actual use of fire by man is in no way natural or positive. Grazing and browsing wild animals probably play an even more important role than does natural fire as an element in the savanna ecosystem.

Grasslands in both temperate and tropical areas represent a wide range of habitats and major regions, determined primarily by rainfall gradient, moisture, temperatures, soils, and herbivorous animals. Some areas are pure grasslands, without trees. Others are wooded savannas and steppes, where trees visually dominate the landscape, but the grasses and bushes still are the most important food resource for the mammals. The soils contain large amounts of humus, much richer than in rain forests. As habitats, grasslands look simple, but in reality they are complex with a number of limiting factors involved. If the ecology of a savanna, a prairie, or a steppe is not studied before man starts to exploit it, the labour may be wasted, and the results of the development scheme prove disastrous. This is a firm rule, but man has not yet even learned that the rule exists.

Whatever grassland habitats are examined—savannas of tropical Africa, flood plains of Asia, steppes of Ukraine, high plateaus of Mongolia, kangaroo grasslands of Australia, llanos, campos, and pampas of South America, or prairies of North America—it is striking how productive they were before man altered them. These grasslands created and supported an array of large mammals, unexcelled numerically anywhere else in the world, which converted green plants into meat in quantities never repeated since. The tremendous productivity was more or less constant and it did not damage the vegetation or the soil. It was a perfect machinery of maximum energy flow, a biological climax, and the highest biomass that each of these regions had known during recent millennia. These optimal conditions of life deteriorated rapidly when man interfered.

Against this background of evidence clearly indicating the beneficent ecological relationships and potentials inherent in natural grasslands, it is tragic to find that man all over the world—particularly European man—has failed to understand, retain, and manage what was good for grassland areas. Instead he changed the nature of the land, exterminated and plundered, introduced exotics, overcultivated and overstocked without regard for the carrying capacity of the environment, of which he himself was and still is just a part.

It would lead too far afield to review here all the errors and crimes man is committing against the grasslands. Agriculture and livestock production seldom has been planned. In area after area, century after century, man has moved into virgin land, opened up the country, and despoiled the vegetation, which rapidly led to a ruining of soil, water, and animal resources. This sad performance continues in an era when the main problem for humanity is the lack of proteins. It will take time to stop the actual destruction of grasslands, and it will take even longer time to reclaim what has been lost. To restore many maltreated areas the cost will be so large and the time involved so great that no politician in one single, short lifespan will find it worthwhile to take action.

Many other areas are lost forever, at least as long as the present climates prevail. But in relation to the current expenses for military purposes or for space research, the amount required to save the world's natural resources does not appear to be astronomically high. Is it beyond reality to hope that man will gradually realize that rehabilitation of what he has destroyed in nature is a condition for survival? This is a matter of education at all levels. Man must learn the consequences of not thinking and planning ecologically. Former grasslands and forests provide grim object lessons.

It may appear to the reader that this book advocates a philosophy based on the assumption that pristine or primeval conditions in nature always are the most productive and that they should therefore suit modern man. While that is not the intention, it does contend that an ecological land survey is a vital necessity before making any decisions on land use and development. The long-range maximum sustained yield of a given area must be established, and a development program directed toward obtaining that yield. When seeking the answer for an area already exploited and therefore probably already destroyed, it is important to go back and retrace the area's former characteristics and the dimensions of its former productivity. Without doing so, its ecological potential will be unknown, and accurate conclusions cannot be drawn.

It is informative to recall what has happened to the North American grasslands. Many of them have been transformed into agricultural deserts by erosion that followed cultivation practices, performed without any regard for environmental conditions. In addition, the cattlemen who invaded the western grasslands after the bison slaughters in the 1870's and 1880's made serious mistakes of overstocking. As early as 1885–87, the combination of overstocking and drought depleted the grasslands so much that about 90 percent of the livestock was lost by starvation. The ground cover was badly damaged by overgrazing and overtrampling. The cattle-raising industry collapsed, but many cattlemen remained in the area. In this situation, farmers invaded the Great Plains, encouraged by the homestead laws. The plows and cultivation damaged the grasslands even more, while the overgrazing effects worsened. The deterioration progressed, and after a few years the farmers gave up. The scars of their mismanagement are still to be seen. The natural regeneration of the Great Plains' grass fields is very slow, especially when they are being grazed and trampled by cattle.

This misuse caused a serious depletion of most grasslands in the range country. During the 1940's, 55 percent of the range area was estimated to be so depleted as to have less than half its original grazing capacity. Another 30 percent was not so seriously depleted, but the forage was far less than the normal grazing value. An overall estimate gave 93 percent of the total range area as being depleted to some extent.

Intensified research coupled with conservation campaigns has led to an amelioration of rangeland and cropland conditions in the United States, but still the western plains are far from being as productive in proteins as they were a hundred years ago, when the white pioneers advanced westward. Today many ranchers on the Great Plains have realized the dangers of plowing up the short grass prairie. They now understand that the economy of grasslands has permanent values that do not exist in one-crop agriculture, with its economic ups and downs and vulnerability to parasites. After a hundred years of hard lessons, recognition has finally been given to the place of grass in the grasslands. Large areas of agricultural land have been converted back to grass country chiefly in the form of cultivated meadows and permanent pastures. This reconversion of plowlands to pastures is good conservation, a step to restoration. An important animal industry can be built up if it avoids earlier mistakes. A complete restoration of the former tremendous productivity of the Great Plains would include the bison. This species is less manageable than cattle, but there is space for both.

Grasslands and savannas of other continents have likewise deteriorated greatly through cultivation and pasture. South America, Africa, Asia, and Australia have seen little or no effort to manage the grasslands, to let them periodically rest and recover. Lack of rotational grazing combined with overstocking has led to decreasing productivity, erosion, sinking water table, drying up of springs, wells, and rivers, and growth of man-made deserts. Shifting cultivation, practiced in tropical forest regions, is also applied in some measure in tropical grasslands. It may result in improvement, but may also have just the opposite effect, depending on local conditions.

In Africa these serious problems face practically all countries—particularly Ethiopia, Somalia, Kenya, Tanzania, Zambia, Rhodesia, Malawi, Botswana, South-West Africa, and South Africa as well as those countries surrounding the Sahara. Probably more than half of the land surface of Africa carries a vegetation in which grass is a dominant or essential element. Grass is the basis for an extremely diversified fauna of large mammals, which convert the grass cover into high quality meat. Livestock grazing in Africa is provided almost entirely by the natural vegetation. Hence the grass cover of Africa is a natural resource of great importance.

A misfortune that Africa has suffered for more than one hundred years is the assumption on the part of former European rulers and those in charge of present technical aid schemes that successful agricultural and stock-raising practices in their temperate homelands are also the best forms of land use for African soils and vegetation. This mistake has led to disastrous failures, irreversibly ruining enormous areas of the continent, and incurring the loss of great sums of investment capital. The most spectacular fiasco was the British groundnut scheme in Tanzania, Zambia, and Kenya, carried out under the Labour Government. Based on a hurried survey—much of it from the air—in 1947 without participation of ecologists, it recommended a huge scheme of clearing 3,210,000 acres of bush for large-scale groundnut and sunflower growing. Haste ruled the decisions and the planning as well as the execution. The crops were ridiculously small in relation to the prodigious costs and gigantic clearing efforts. In 1951 the scheme was abandoned, and £36 million English were written off by the British government. This sum does not include the lost values of ruined lands that intact with a rich wildlife were much more productive before the bulldozing and ploughing destroyed the climax habitats they contained. Untouched African savannas and bushlands have a remarkable array of plants and animals that have brought productivity to the highest degree the habitat is capable of. For man, it is more economic to cooperate with these climax biocommunities than to smash them in the hope that their harvest will be greater after performing a major alteration of the environment. Calculated short-term profits by exploitation of soil and plant resources are often advantageous for a few people, but the long-term benefits on a sustained yield basis are lost to society.

Very few, if any, savannas in Africa have not been disturbed by various man-made factors such as fire, overgrazing by livestock, clearings for cultivation, and so forth. In particular, grazing and trampling by large concentrations of livestock have caused degradation of the vegetation and erosion over extensive areas. This is not the case when the savannas are grazed by wild ungulates. The impressive host of African grazers and browsers represents a spectrum of selective feeders that are in ecological balance both with the vegetation and among themselves. It is a kind of collective symbiosis, a partnership between the vegetation and the animals wherein each element benefits the other. Natural selection has adopted each species, plant or animal, to occupy a particular ecological niche. The animal productivity increases with the number of species available to utilize the savanna vegetation properly —that is, when each ecological niche is filled. When only some niches are occupied, the productivity decreases. The complex relationships between vegetation and wild herbivores are upset when the latter are replaced by grazing domestic animals, since the grazing pressure of cattle, goats, and sheep is much greater both in time and space (see page 203). The result is a much lower biomass per unit of area, a general degradation of the environment, and often a collapse of the whole ecosystem. This is what has happened over large areas of Africa. Many of the lands devastated by cultivation and overgrazing have had to be planted with grass to enable them to recover. But the grasslands need the age-old partnership with herbivorous animals to regain their ancient richness.

In Australia, grasslands were grazed in a virgin state by wild mammals and birds for thousands of years and no erosion occurred, but when Europeans pastured sheep and cattle there, disastrous effects soon followed. Degenerated pastures and removal of surface soil by erosion now characterize all Australian grassland communities.

In the Asian tropics and subtropics, almost all existing grasslands have deteriorated seriously, with destruction of soils. In Pakistan, India and Ceylon most of the existing grasslands are man-made. Man has likewise greatly modified the South American grasslands, which have lost much of their wildlife.

All over the world large grassland areas are marginal in the sense that they are suited neither for cultivation nor for livestock pasturage. This is notably the case in Africa, where about 90 percent of tropical lowland savannas are in the marginal category. The best use of such lands would be to utilize rationally what they produce in their natural state. Their productivity is often highly important. This is discussed in the chapter on the animals (see pages 201–204).

The simple fact is that large areas of the world's rangelands are overused. Many of them are continuously deteriorating due to overgrazing and overtrampling, and cannot sustain sedentary or seminomadic pastoralism. The time is at hand to consider a global plan of grassland restoration, before deterioration has gone beyond the point of no return. Long-term programs are needed for repairing and rebuilding lands that have been seriously damaged by the mistakes of the past. They will be costly and will require cooperation among nations. In addition to resulting in higher quality and quantity of resources produced, such programs may also lead to a better way of human life—one in harmony and cooperation with the environment.

Forests

"Any fool can destroy trees," said John Muir, a great American conservationist. "They cannot defend themselves or run away." Man has foolishly destroyed trees all over the world on a gigantic scale. The destruction is truly foolish—the same story, already so often encountered in this book, of reckless destructive exploitations without regard for replacement or for management to prevent further depletion. After waves of destructions, the once well-balanced forest lands were left fully exposed to wind and water. Whole regions broke down because the ecosystems collapsed, and in time nature struck back with floods and other catastrophes. Human populations paid tribute in suffering and economic loss, but those who were hurt were not necessarily the same ones that had profited from the destruction. In many areas of the world the damage that grew out of heedless, wholesale logging must have cancelled the revenue or turned it to an economic loss.

Profligate forest exploitations do not just belong to the distant past. In the 1950's man ruthlessly carried out uncontrolled destructive logging on the Douglas fir forests of northwestern California. Nature responded with two flood catastrophes of tremendous dimensions in 1955–56 and 1964. The floods covered valleys with silt and debris from the mountain slopes. Entire human communities disappeared, almost without a trace of their former existence.

This California example is all the more impressive for coming from the United States, where governmental conservation is stronger than anywhere else in the world. It indicates the scale on which forest destruction is proceeding elsewhere, in countries less advanced in conservation. In fairness, despite a century's incredible man-made forest devastation in California, it must be recognized that this state still has one of the largest areas of old-growth virgin forests in the United States.

The forest's earth coverage is decreasing rapidly, particularly in the tropics. At present about 9,800,000,000 acres of the earth's surface is covered by forests. Only about 13 percent is under some form of management, not necessarily a rational utilization.

In evaluating forest resources one must count much more than just the timber value. The total ecological role of forests must be taken into account. Too often the physical importance of forests as climatic, hydrographic, and ecological stabilizers in ecosystems is either overlooked, underestimated, or simply neglected. These forest roles are essential for environmental health in all climatic regions, especially in the tropics.

Among the many beneficial environmental functions of forests, the following are globally important:

1. The climatic role, by assimilating carbon dioxide, purifying the atmosphere, and producing oxygen; providing areas of low re-

flectivity and high heat absorption with low heat conductivity; intercepting moist air and inducing precipitation; acting as windbreaks and creating aerodynamic roughness.

2. The hydrological role, by serving as watersheds; by accumulating, cleaning, regulating, and distributing water resources; by preventing silting of lakes and dams.
3. The ecological role, by preserving and forming soils, by maintaining the basic elements for the preservation of habitats; by producing food and habitats for wild animals.
4. The economic role, by producing timber, wood for fuel, and chemicals; by providing recreational facilities; by functioning as shelter belts, thus increasing qualitatively and quantitatively crops and domestic animals; by reducing noise pollution.

All these beneficial effects of forests emphasize the importance of considering carefully all deforestation programs and of doing it in a wide perspective by including all ecological factors involved.

Deforestation is usually accompanied by adverse environmental effects of which the most important are:

1. Changes of microclimate and climate (depending on the size of the deforested area).
2. Changes in hydrological cycles and water regimes, resulting in erosion of river banks, sedimentation in rivers, lakes, and seas, adverse flooding of often catastrophic character; perennial streams and rivers become seasonal or dry up entirely; decline in water quality; collapse or adverse changes of aquatic ecosystems in rivers, marshes, and lakes with negative effects on fisheries.
3. Decline or local extinction of wild terrestrial animals.

The disadvantages of deforestation are ecologically and economically manifold. On the other hand, there is a growing demand for timber and forest products as well as for clearing forests for so-called "land development" for industrial, arable, and pastoral purposes. It is estimated that by 1985 there will be a doubling of demand for timber.

Forests occur from subarctic regions to the tropics and provide a wide range of habitats. The most fully developed, but perhaps the most vulnerable, are the tropical rain forests, both montane and lowland. They are the greatest biological achievement existing in our time. The wealth of species is overwhelming. In Indonesia, exclusive of Java, about 3,000 species of trees are growing; in the Malay Peninsula, about 2,500; in the Amazon, at least 2,500. The relatively small area of the Ivory Coast of West Africa has more than 600 species. As would be expected, the diversity of shrubs, herbs, lianas, and epiphytes is much greater still. The phanerogamic flora of Malaysia and New Guinea totals about 20,000 species.

The highly complex rain forest ecosystem has existed for millions of years, and virtually only within the last hundred years has man disturbed it. Human populations living in the rain forest—as do the Congo Pygmies, for example—do not destroy it. They utilize the forests like all other animal species—by utilizing the surplus. As food gatherer and predator, the Pygmy is part of the environment. As long as he does not cultivate, this harmonious relationship with the habitat will remain.

A present danger in tropical land use is the exploitation of rain forests for timber and shifting cultivation. Within a brief span of years, it is possible to see dense forest areas converted into open country, which soon deteriorates. When this happens, the soil may suffer a permanent change, and the habitat may enter a phase of irreversible destruction. In Asia, Africa, and South America, man has cleared enormous areas for cultivation, plantations, and settlements, and this activity is accelerating. If the rate of destruction continues, one of the world's most spectacular expressions of natural wonder is doomed. But it is not just for their sheer magnificence that rain forests should be preserved. They are useful as well —though little is as yet known about the usefulness of the vast variety of tree species that grow exclusively in these forests. We know the timber value of a dozen kinds of rain forest trees, but virtually nothing of the biochemical values of many thousands of species. Moreover, the rain forests are the home of a multitude of animals that do not exist in any other environment. Some of these animals have been found to be extremely useful to man, and many more may yet prove so. What will be the effects on climate if the rain forests vanish? What will be the consequences for the soils? The montane rain forests regulate much of the hydrography of the mountain slopes and regions below. To destroy these forests would be one more invitation to erosion, floods, landslides, and a gradual decline of agricultural lands through desiccation and sinking water table. The rain forests also serve as reservoirs of genetic material. Providing habitats for a tremendous number of species and genera, the rain forests have for eons been a center of evolution, speciation, and dispersal. Much of the temperate flora, for example, derives from the tropics.

The rain forests of Africa represent one-third of the present rain forested areas of the globe. In 1959 they occupied 1,300,000,000 acres of their original range of 3,734,900,000 acres. The figures dramatize the alarming rapidity at which the rain forests are disappearing. In southeastern Asia the process is even faster. In view of the tremendous ecological role and multiple values of tropical forests (lowland and mountainous) in a, for human beings, long-term positive sense, it is highly regrettable that FAO and other aid agencies, as well as former colonial forest services operating in developing countries in Africa, Asia, and other tropics, almost indiscriminately for decades have recommended getting rid of the native forests in order to replace them either with exotic trees like pines or eucalypts or by short-lived cultivations or pastures. This policy of regarding natural tropical forests as waste products is dangerous and accelerates the deterioration of originally productive land. It has unnecessarily destroyed enormous areas of useful forests and unique samples of national heritages.

This process of destruction is still going on on a big scale. Fortunately, there is a growing reaction among Africans and Asians to the destruction of indigenous forests in these countries. If humanity destroys the tropical rain forests, it will be an economic and ecological tragedy. Ethically the crime—to put it quite plainly—would be uncivilized.

Shifting Cultivation

As noted above, shifting cultivation (*ladang* in parts of Asia and New Guinea, *taunggya* in Burma, *togel* in Java, *ray* in Cambodia, Laos, and Vietnam, *kaingin* in the Philippines, *chitenene* in Africa, and *milpa* in parts of central America and Mexico) is the major threat to tropical rain forests everywhere, and in the present day particularly in Africa and Asia. It is a traditional slash and burn method of agriculture, perhaps the oldest one, presumably existing since the Stone Age.

The cultivation methods on cleared farms in the tropical world seem to vary little and are quite simple. They may be observed wherever one finds the shifting cultivation farmer at work. In the Congo the farmers employ local minor variations from the basic simplicity that involve a degree of extra effort, such as the use of collected animal manures and additional ash fertilizers. However, the clearing methods seem to follow the same pattern in all tropical forests, with only minor differences. The periods of crop growing and fallow vary locally. The best way to explain shifting cultivation is to describe a typical cycle as it results from man's activity followed by his abandonment of the area.

The technique of preparing a rain forest for cultivation is usually to clear an area by cutting small trees and shrubs, which are left to dry at the base of larger, ring-barked trees. At a late stage of the dry season this debris is set afire. The flames kill many of the larger trees that have a thin bark. In some regions dried lianas are used to put fire to the leafy crowns of the dead trees, so that they also burn from the top. While the ashes fertilize the surprisingly thin humus layer on which crops are grown, the fire destroys many of the organisms in the topsoil. The structure of its organic matter totally changed, no longer nourished by the rain forest debris, and exposed to sun, wind, and rain, the soil rapidly loses its fertility. After a few years, when the farmer no longer can subsist on his plot, he abandons the farm. He repeats the process in a neighboring area. So it continues, over and over.

When a farm is abandoned, the open soil is invaded by herbs, bushes, and climbers, which soon cover the whole area. Among them are some trees, which due to efficient seed dispersal and other characteristics rapidly colonize cleared areas. After about three to five years, an almost impenetrable tangle of vegetation covers the area of the abandoned farm. At this stage a farmer often comes back for a second clearing and farming of the area. This time the fertile period of the soil is even shorter, and crops are grown only for two years or so.

If the farmer does not return, and nature is left to itself, the forest recolonization proceeds. Most of the pioneer trees have a rather short life span—about 30 years. They are gradually overtaken in size by long-living species. The shadow of these disfavours herbs, climbers, bushes, and smaller trees. The community shifts to a real forest, though at this stage still of a secondary character. After about 75 to 100 years such secondary forests come to resemble primary ones, but they still have features that indicate their relative youth. After about 250 years, secondary growths can scarcely be distinguished from true virgin forests.

As long as the human populations of the tropics were small, the practice of shifting cultivation did not constitute a danger to the rain forests, which regenerated faster than they were destroyed. Today the situation is quite different. Increased population has put additional demands upon the land, so due to population pressures, the fallow periods are short, and the forests have no chance to recolonize cleared areas. The soils become irreversibly impoverished and damaged. As a result, in Asia, Africa, and Latin America shifting cultivation is causing the rain forests to disappear at an alarming rate.

Another menace associated with shifting cultivation in many areas of tropical Asia is the spread of cogon grass (*Imperata*). This is a stiff grass, generally unpalatable to domestic and wild herbivores, that rapidly colonizes frequently burnt, cleared, and cultivated lands of reduced soil fertility. In Thailand it is actually called *lalang* grass, which means shifting cultivation grass. It burns rapidly, and the traditional procedure is to burn over the cogon areas in the dry season. This prevents forest regrowth and makes the grass even denser, transforming the area into land that is completely useless for both animals and man. This devastation has laid waste large areas of southeastern Asia, particularly in Thailand, Cambodia, Vietnam, and the Philippines. It continues on an accelerated scale. If not stopped, the practice will ruin these countries by making permanent cultivation, sylviculture, and game meat production impossible, and by destroying watersheds.

The rapid loss of fertility in soils where annual crops are grown could perhaps be counterbalanced to a degree by the use of chemical fertilizers, but rapid leaching due to heavy rains often reduces the effect of fertilizer. Moreover, they are seldom employed by farmers in tropical forests, probably because they are difficult to purchase, too expensive, or simply unknown. Where the natural forest is cleared, and other tree crops such as cocoa, coffee, oil palms, or rubber trees are substituted, the soil is less damaged, and fallows are generally not necessary. Like the wild trees and shrubs, these cultivated species can capture soil nutrients and even nourish the soil layer. Permanent cultivation of such tree crops are practicable, which explains their success in several tropical countries.

Shifting cultivation in the tropics is an extremely wasteful exploitation of a useful natural resource. For a nation it means tremendous economic losses of forest, soil, and watershed and wildlife resources. Since the harvests from areas cleared by shifting cultivation are progressively reduced beyond subsistence level, the situation for the farmer is inherently precarious and will in the long run become impossible. Continuously destroying his own habitat, he is doomed to perish or to change environment. He not only jeopardizes his own future but that of whole regions and in many parts of the world entire nations. To reduce the amount of food available to communities already existing at a minimum physiological level is to reduce the population. Shifting cultivation benefits nobody except temporarily and then at a price to country and society too great to estimate. In the Philippines, where the forest resources are being destroyed at a rate hard to believe, an official committee states that in forest products alone the financial loss to the country from the 13½ million acres of cleared forest land is about thirty-three billion pesos, or five billion dollars.

Some 200 million people presently practice shifting cultivation on about fourteen million square miles of the global surface—more in tropical Africa than in Asia or South America. The survival of tropical and subtropical countries and their populations requires that they stop this self-destructive form of agriculture immediately. It should be replaced by a sustained, permanent type of cultivation, with crop rotation, on suitable lands. Sound conservation and utilization of the forests would contribute positively to the economy of each country and its people, and the whole world would benefit.

Forest and Water Economy

Modern forest conservation and management are designed to produce not only timber and wood products but also water, soil, and animals. Water is especially important, because it affects the productivity and well-being of areas located far from the forest source. Forests keep soil erosion from wind and water runoff at a minimum. They capture large water volumes coming from irregular rainfall, and their shade retards the spring melting of the winter snows. They store water for long periods and supply it to surrounding lands throughout the year. At the same time the forests serve as a fishing and hunting ground. The principle of forest multiple use has great importance for the social economy and prevailing living standards.

The water storage capacity of forests and their role as regulators of water supplies are not always fully realized. In the United States about one-half of the nation's stream flow comes from forests. In the western part of North America as much as 90 percent of the usable water yield originates on forested watersheds. About 75 percent of the total forest area of the United States is needed for surface water supply and groundwater recharge as well as for control of floods. In tropical and subtropical regions especially, the forest areas must be preserved to maintain the hydrographic balance of a region, to keep the rivers flowing in the dry seasons, and to stabilize the soil. This is all the more true for montane forests, because a major part of the sustained fertility of surrounding lowlands depends on them. But these forest-water relationships hold in temperate regions as well.

In a living, productive landscape, natural or cultivated, effective watershed conservation is easy as long as the forests remain. Dense, undamaged grassland can likewise be good watershed. But it will be difficult and costly to restore bare, eroded hillsides where water rushes down during the rains.

Prior to exploiting forests, it is most important that a detailed ecological survey investigate the watershed situation and its bearing on the ecosystem, so that forest logging activities will not ruin agricultural lands in the area or the water supply of a town situated miles away. Actions taken without full knowledge of ecological relationships are dangerous to man wherever they happen, but they are especially risky in the case of forests. The wooded habitats interplay with the living landscape or ecosystem in a more complex way than any other major biome. For the survival of a living landscape in a broad sense, the preservation of forests is a necessity.

Multiple Use of Grasslands and Forests

The vegetation resources of the world will be wasted, and with them many other renewable natural resources, if the organizations and people involved in the utilization of grasslands and forests do not organize and coordinate their activities. Conservation and management policy should be set by government, and measures should be taken to assure that the ecology of the regions and their resources are well understood. If all this is accomplished, a rational utilization of the vegetation in most cases inevitably leads to a multiple use of grasslands and forest. Any major natural resource—a savanna, a forest, or a region consisting of several biomes—may serve many purposes of value to society. Without ecological surveys and proper land use planning, it is hardly possible to determine what the multiple uses should be, and just how they should be coordinated. Conflicts between useful functions of land often arise. When this happens, priorities must be evaluated relative to the sustained carrying capacity of the area and the needs of the country.

What will assuredly emerge from such planning is a realization that there are many "multiple use factors" that warrant retaining grassland and forest habitats instead of destroying them for other uses of land. Some of these are timber, wood and other plant products, wild and domestic animal fodder, gene banks for plants and animals, shelter and habitats for the fauna, soil stability, water maintenance and regulation, climatic influence, oxygen production, productivity samples of natural habitats, sites for basic research, human education and recreation, and so forth.

It is remarkable that all over the world—at least this is my experience—so few foresters are conservationists. They usually believe they are, but the average forester does not realize the multiple values and the general ecological role of forests. For a forester conservation generally means that he can extract a certain amount of timber from a forest or convert a natural forest to a uniform one for later timber production, or that the timber outtake does not exceed the growth of replacement. At the present time only about 10 percent of the tree species of tropical rain forests are of commercial value, that is, timber value. The other 90 percent are often regarded as waste. They may be of great value in the world of tomorrow and are certainly of ecological value today.

Since so few species of trees in relation to the total number of species occurring in tropical rain forests are at present commercially valuable, the exploitation should concentrate on them and leave the others intact to preserve the habitat and its animals. The removal of such mature trees of certain species in a cycle which allows a regeneration period of forty-five to sixty years should not disturb the forest as a biome, provided the logging is done wisely. However, where there is any cutting system employed at all, the cycle is usually much shorter, with a cut every ten to twenty-five years. In the the long run such a short cycle is not ecologically sound and gradually leads to habitat changes and usually to a reduction of the multiple use values.

Productivity and Biomass

Grassland and forest ecosystems have a high biological productivity. A number of methods may be used to measure the productivity of an ecosystem, a biome, a habitat, or a small site. The method may be based upon various components living in the area concerned: species, or groups of species, of plants and animals, or the total of both. Or it may be expressed in other terms: energy conversion rate, standing crop biomass, annual harvest weights, and so on.

In this connection it is interesting to refer to estimates of the U. S. Department of Agriculture that two million acres of land each year, excluding surface mining, are converted to nonagricultural use. Half of this is shifted to such uses as wildlife refuges, recreation areas, and parks. The other one million acres are converted to more intensive uses. Of that, 160,000 are covered by highways and airports. An estimated 420,000 acres become reservoirs and flood control projects. The remaining 420,000 acres are developed for urban uses. About half the national acreage converted every year to urban uses is cropland and grassland pasture. The rest is forest and other land.

In the long history of pastoralism and agriculture man and society seem to have paid little attention to the potential sustained average yield per acre. In any event, the productivity curve went steadily downward during long periods until the land became exhausted and the economy collapsed—and empires fell. In recent times man has become aware of the carrying capacity and productivity potentials of various lands, but he makes use of his newly acquired knowledge in only a few areas. The fact that the more productive crops increase food supply, providing the land can continuously yield maximum quantities without deteriorating, is often forgotten due to ecological ignorance and short-term planning. At the same time the usefulness of natural habitats and their productivity have been neglected.

The planning of efficient utilization of vegetation resources must take into account the sustained productivity, biomass, and carrying capacity of an area under entirely natural conditions by comparing their parameters with the estimated long-term productivity of the same land under cultivation. Records for already exploited lands, which once carried similar habitats, are usually available for comparison.

Contemporary man must avoid falling into the same trap as his ancestors, who failed to understand the complex relationships among vegetation, soil, and water, wherein carrying capacity and productivity are significant details. He must learn to utilize renewable natural resources without damaging them. An "ecological way" of land use will result in great public benefit. Natural productivity and standing crop biomass are useful tools that should be utilized in all land use planning.

Deforestation

In wooded countries land clearing has proceeded, with few exceptions, almost as long as man has inhabited them. The magnitude of the resulting environmental change is best illustrated by figures. It is estimated that the area of virgin forest within the present boundaries of the United States at the time Columbus came to America was approximately 800 million acres. By the end of the 1920's it was less than 100 million acres. Today it is 56 million acres—7 percent of the original area. When Europeans colonized the United States, about 45 percent of the land was covered by forests. At present it is about 20 percent, partly thanks to energetic reforestation.

Even if the forest destruction in North America appears frightening, especially in view of the short period involved, it is nothing compared with what has happened in many other areas of the world. Spain was once covered by forests, but now only one-eighth of its territory is forested. In Greece during historic time, 60 percent of the land was covered by forests. Now the figure is down to 15 percent. Nigeria, with one of Africa's largest populations, has eliminated 74 percent of its forests. As noted earlier, Africa's rain forests presently occupy only one-third of their original area. Only 9 percent cf China's once heavily forested land still carries forests. In India the corresponding figure is 18 percent. In Ceylon, only patches remain of a forest that 110 years ago covered most of the island.

The Philippines have lost 50 percent of their forests during the last fifty years alone, and 80 percent of this loss was cut during the last twenty years. And only one-sixth of the total cut has been used for timber. The rest has been burnt or left to rot. As a consequence of this heavy forest destruction, one-fifth of the Philippines' land area is now totally unproductive. Of Brazil's immense forests, about 40 percent was estimated to have been cleared by 1955. In Puerto Rico—just one example from the West Indies—one percent of the virgin forest remains. Only 2 percent of Australia is forested land. New Zealand was covered by about 62.5 million acres of forests at the beginning of European colonization about a hundred years ago; now there are only 14.5 million acres, a large part of which consists of exotic species planted by man.

In Africa the greatest threat to montane forest areas, after shifting cultivation, is from excessive numbers of livestock in the lowland areas. The domestic animals are deliberately driven into the montane forests. Cattle and goats eat and trample down the young shoots of trees. The animal husbandmen slash and burn to open up new pastures, and these creep higher and higher up the slopes.

Wood-cutting for domestic fuel is a serious problem in many areas, particularly in arid and semi-arid regions. The woods at the fringe of deserts and semideserts disappear very rapidly and give way to desertification. It is estimated, for example, that in Kano in Northern Nigeria nearly three-quarters of the firewood consumption of some 75,000 tons per year, for a population of about 300,000 people, is brought in from within a radius of about twenty kilometers. This represents the felling of tens of thousands of trees every year, and there is hardly any regeneration because of overgrazing by goats and cattle. Charcoal is used industrially in many parts of Africa, and this contributes to a rapid disappearance of savanna woodlands.

Forests persist mainly in those areas of the world that are unsuited for agriculture or other development. But even within the forested areas many virgin or indigenous tree species must give way to exotic species that are considered to be more economic, faster growing trees. In this way even some excellent timber trees have become rare. Many fast-growing species produce less valuable timber than what is furnished by those that grow slowly. Some species of mahogany from the West Indies and teak from southwestern Asia, for example, have become scarce on the world market. Trees of quality are sacrificed for trees representing quantity.

The reduction of forests all over the world eliminates the most complex habitats, which are extremely rich in biotic elements and communities. A biological capital of immense value is replaced by uniform habitats. The long-range values of these simplified biotic environments are unknown, though man, blinded by short-term advantages, refuses to recognize this uncomfortable fact.

Gallery forests, or riparian forests surrounding river courses, are often highly productive and, particularly in the tropics, provide environments for the existence of many aquatic life forms as well as for terrestrial animals. For example, an astonishingly high proportion of the numerous species of fish occurring in the rivers of the Amazon basin feed on organisms that are produced by the terrestrial environment, chiefly from the surrounding shore forest. The aquatic biological cycle of the Amazon system itself is rather poor in nutrients. Many species of Amazon fish feed directly on leaves, seeds, fruit, and terrestrial invertebrates that are dropped from the riparian vegetation. Since it is essentially the tropical forests surrounding the rivers that maintain the extremely rich fish fauna of the Amazon system at its present high level, it is probably the same for rivers in the rain forest regions of Africa, Asia, and Australasia. The accelerating destruction of those forests directly threatens the aquatic life, including the predatory fish that form part of the food chain, and may lead to its disappearance. Thus valuable aquatic food resources for the human populations will be destroyed along with the forest—an instructive example of the complex ecology of aquatic-terrestrial communities.

Also elsewhere gallery forests are at present destroyed very rapidly. In Africa, riparian forests have been reduced in tsetse control programs. In Asia they are partly destroyed by irrigation and drainage schemes.

Burning

Grasslands and forests in many parts of the world are exposed to man-made fires. The purpose of burning differs widely in different regions. Nowhere is the practice of burning so widespread as in the tropics, where it is an important factor modifying landscapes. In Africa, bush fires—burning of grasslands—are even more frequent than burning of forests. In fact, fires greatly influence the vegetation of African savannas. Cultivated lands are also burned regularly. As one flies over Africa, smoke from numerous bush fires rising toward the sky or spreading over the plains is a common sight that continues for hours. By night the fires illuminate the countryside. Farmers start bush fires regularly in order to burn off dry, old grass, to produce ashes valued for the nutrients they add to the soil, to kill weed seeds, or simply to clear the savanna for planting crops. Livestock owners burn dry grasses so that fresh new grass will shoot up in the ashes, providing pasture for cattle, or to cover their tracks when stealing cattle. Hunters also burn the dry grass because the new green shoots attract grazing antelopes, which make easy targets for snaring. Or the hunters start fires in order to drive out animals for hunting. Honey collectors, or honey hunters as they are called in Africa, regularly set fires even in areas not inhabited by man. They use fire as a tool for collecting honey and just leave the fire burning after them.

African savannas are usually burned once a year, in some areas twice. In rare instances some parts may escape burning for a year or two. The regular burning produces a kind of fire climax vegetation, where fire-resistant, thick-barked trees, bushes, and perennial grasses dominate.

The ecological effects of burning on the African savannas are a matter of controversy. Normally fires maintain savannas, but even without burning there would certainly still be savannas because grazing animals leave an impact in that direction. Fires may be useful as a management tool locally, and occasionally their short-term effect on the vegetation may also be beneficial. But in the long run, annual burning is in general detrimental to the soil, the vegetation, and the animals. It reduces the nitrogen content and the organic matter in the topsoil, impoverishes the vegetation, and destroys or wipes out a number of animals in and on the soil. It also leaves many animals without food—browsing mammals, for example. In short, annual indiscriminate burning slowly reduces the general fertility of the area. The controversy about the effects of burning tropical grasslands is brought about by the lack of knowledge of fire ecology. This could be dispelled by getting information on this important subject through well-planned and executed research programs. This would need to be carried out on all the various vegetation types over a period of years.

Europeans as well as Africans often claim that burning is necessary to remove dry grass and help fresh grass to grow. But only seldom does dead grass inhibit the growth of its offspring. In reality, many factors other than fire destroy the mat of dead grass. It is trampled and crushed by many animals, eaten by rodents and termites. It decomposes, contributing to the humus like all organic litter. As a part of the soil formation it is, like dead trees, host to numerous tiny animals living on or inside the dead stems. The raging fires that annually strike African grasslands deprive the savanna habitats of many life forms that could have existed there, and kill so many other organisms that have colonized the area since the previous fire, that the whole situation must be regarded as unnatural—even if the procedure has been followed for centuries. Man has become much more numerous in Africa during recent decades, and he continues to burn as did his ancestors. Gradually larger and larger areas are affected by the fires. Moreover, the fires are not as a rule controlled. They often occur at the wrong time of the dry season and of the day, and they destroy indiscriminately.

Ecologically and biologically, annual burning is a waste of resources, a destruction of organic matter in many forms—decomposing material, stored seeds, eggs and pupae, growing plants and animals. Fires prevent diversity of biocommunities and habitats. At the least, large areas in nature reserves should in principle never be burned artificially, but rather should be protected from fires moving in from outside areas. Planned fires can stop accidental fires. Firebreaks can do that job too, but they are expensive. Moreover, regular burning at the edge of savannas toward semideserts and deserts enables the latter to advance and conquer productive grasslands. Similarly, fires sweeping through the edges of forests force them to give way to savannas. Fires in bushlands and thickets likewise create savannas.

In Madagascar 90 percent of the fires that every year sweep the country are set by cattle owners. Unfortunately they don't care if the fires spread uncontrolled over enormous areas and burn the few remaining natural forests and even planted forests. The burning in Madagascar is the prime cause of the country's denudation and the rapid degradation of its natural resources. The repeated fires, combined with the trampling by cattle, have made the soil sterile and skeletonlike over very large areas.

Among the great national parks in Africa, probably the Kruger National Park of South Africa has the greatest long-term experience with scientific study of fires as a management tool. In 1968 a symposium on fire problems was held there. Out of the discussions of the pros and cons of burning came a series of key points. The cons may be summarized as follows:

1. Vertebrate and invertebrate animal populations are adversely affected. Elephants have been killed in bad conflagrations. Even in slower, cooler fires, tortoises, snakes, mice, and insects are exterminated, bird eggs are burned, the habitats of smaller animals are destroyed, and food supplies of all species are laid waste, if only temporarily.

2. Destruction of food supplies, especially under unfavorable conditions, results in abnormal concentrations of animals in the remaining undamaged areas, producing competition that may lead to the death of weaker animals and the destruction of vegetation.

3. Exposure of the soil surface, i.e., devastation of essential organic material, may result in heightened loss of soil moisture, overheating of the soil surface by the sun, total or partial ruin—depending on the intensity of the fire's heat—of the microfauna and flora in the upper soil layers, erosion by wind and rain, and greater runoff of water. Together these consequences could be held coresponsible for drying up of wells and sinking of the water table level.

4. Some plant species may be changed structurally, or even be totally destroyed by fire, with the result that the physiognomy of a plant community may change entirely.

5. Reproduction may be retarded or even prevented, especially of those plant species having flowering and seed formation periods that happen to coincide with fires.

6. Apart from the above-mentioned effects, woody plants and even grown trees may be adversely affected by fire-caused wounds. Unsightly scars remain, and these may give access to insects and so lead to gradual decay.

On the other hand, it was pointed out that the existing plant population has reached its present status in spite of a possible excess of burning, and it can be logically accepted that the status quo cannot be maintained without the firing practice. A careful evaluation of the detrimental effect includes consideration of the time of year when the fire occurs as well as its intensity. Most of the harmful effects, it is maintained, disappear completely or are reduced to a minimum when the fire occurs under favourable conditions. In other words, wisely "planned" fires to accomplish specific objectives in various types of grassland are often beneficial, while unplanned fires in almost all instances are detrimental. Also, one may as well face the fact, at least in Africa, that if there are no "planned" fires to prevent unplanned fires, the latter are going to occur. The destructive environmental influence of annual burning cannot be sufficiently emphasized. Planned burning should be regarded as a management tool, and like the plough or atomic energy, it is up to man to use for his benefit and survival. The following points are made in favour of veld burning:

1. The fire removes old, dry litter, so that the danger of devastating accidental fires is minimized, and grazing that might otherwise be practically worthless becomes available to animals in the form of tender, palatable growth.

2. Because of the inherent differences in palatability of various grass species at the mature stage, animals graze selectively. The more palatable species are thus subjected to such concentrated grazing that eventual eradication is possible. On the other hand, all species are palatable enough in the young, succulent stage, so that selectivity plays a lesser role. Thus unpalatable species are temporarily, at least, subjected to intensive grazing, with the possibility that this may aid the better types of grasses to flourish at the expense of the poorer ones.

3. Some plants are vegetatively or reproductively stimulated by fire.

4. Possibly the most important usefulness of fire is to counteract bush encroachment. This claimed advantage is probably the most controversial. One school of thought believes just the opposite—that fire is the actual cause of bush encroachment. Single-stemmed, woody plants, it is claimed, become many-stemmed in regrowth, occupying a larger area and growing bushier than before being burned back. In the absence of fire, this school maintains, the denser grass cover would restrain the existing woody plants through keener competition and in time would dislodge them. Moreover, the establishment of new individuals would be greatly discouraged.

It is impossible to generalize the pros and cons of burning in tropical areas. Each area has to be considered separately, relative to the kind of land use to which it is put. But there is no doubt that for natural communities, repeated artificial burning is detrimental to productivity.

In temperate regions most fires are accidental, but deliberate burning also occurs, chiefly in forestry. Unfortunately, for example, in coniferous forests of Scandinavia and Finland most of the burning that follows logging, to clear cutover areas, takes place in spring over scores of thousands of acres during the breeding season of many ground-nesting birds. Many species among these, like capercaillie and black grouse, are useful game birds. Many mammals, reptiles, and amphibians also succumb, as well as useful ants and other invertebrates.

The moors of Scotland are to a great extent a result of man-made fires. There, burning regularly occurs as a tool to manage pastures for sheep. The effect of heath burning on wildlife is a much discussed subject. Burning may increase the density of breeding red grouse (*Lagopus lagopus*) but not necessarily its breeding success.

Burning of heather (*Calluna vulgaris*) modifies the structure of the heath as well as its productivity and chemical composition. The yield of heather shoots from newly burnt areas is lower than from a mature stand. It may take several years if regeneration is slow before the ground cover of the plant is fully renewed and productivity again attains the maximum. Therefore the increase of red grouse density after burning is not due to an increase of productivity of the heather. In fact, grouse numbers did not increase until three years after burning began (Miller and others, 1970). However, it seems to be still unclear what environmental effects the burning of heather has on the whole complex of soil, flora, and fauna and their many ecological variables.

The opening up of temperate forest areas by access roads and the increase of outdoor recreational activities have greatly increased fires caused by human carelessness. In the United States alone between four and seven million acres of forests are burned every year. In Australia fire has been a major factor in destruction of a large part of the native vegetation during the less than two hundred years of European occupation.

Cultivation and Farming

Man clears, ploughs, and cultivates. He removes the natural vegetation and replaces it by monocultures. The steel in the grass and the forest symbolizes the violence of the machine age. Ploughs, tractors, mechanical saws, and bulldozers turn grasslands and forests upside down, altering them completely and forever. The changes in vegetation during recent centuries alone are of such proportions that they have altered the face of tremendous areas, even entire countries. South Africa is an example. Once the fertile veld of the Karoo was covered with perennial grass, utilized by gigantic herds of grazing ungulates. Today nothing remains of those fantastically rich resources. The Karoo is mostly a subdesert steppe, with partly bare soil dotted with drought-resistant patches of stiff grass, shrub bushes, and acacia thorns. The deserts and semideserts are spreading irresistibly eastward, threatening to engulf the rest of southern Africa. Further north, in semiarid areas adjoining the Kalahari Desert, much the same is happening. This threatening calamity is a consequence of what man calls "cultivation." The grim drama of land deterioration and desert invasion is nature's backlash against man's misuse of the vegetation, his failure to understand the partnership between soil, grass, and wild animals. The destructive forces at work on the veld may eventually be put under control if effective conservation measures are taken. Costly operations have been initiated. The ploughed lands have been planted by grass. Herds of hoofed animals must be brought back, and the attempt is being made with beef and dairy cattle. Whether livestock can reestablish the ancient partnership between grass and ungulates remains to be seen.

This South African example is but one of many showing where cultivation may lead when man forgets to cooperate with nature.

Overgrazing

Overgrazing occurs when grassland habitats are overstocked by cattle, goats, and sheep. When the North American prairies were invaded by livestock, the destruction took place in a temperate region, where the vulnerability was less, and the healing capacity greater than in subtropical and tropical regions. In the latter areas destruction by overgrazing is widespread and goes on continuously. It has far-reaching negative effects on the protein productivity and the national economy of many countries. Evidences of this disastrous situation are manifold. The subtropical lands of the Mediterranean, western Asia, and northern Africa bear terrible scars from overgrazing. In the tropics the effect may be even worse.

Overgrazing seldom occurs where natural populations of animals utilize natural grasslands and woodlands—that is, where predators control the herbivores. In artificial habitats, where monocultures and other cultivation practices have substituted crops in place of useful food plants for herbivores, many animals are forced to feed on the cultivated plants.

In Africa a host of wild grazers and browsers utilize the savannas by feeding upon different vegetative levels, the ecological niche preferences of different species making for efficient use of the savanna plants. They feed on grass and woody plants ranging from low herbs to tall bushes and trees. Some even consume thorny twigs and branches, but grass is the preferred forage for most species. African grasslands grow more than a hundred species of grass and associated herbaceous plants. Each one is a part of an ecological niche. On the animal side, each species of the array of grazing ungulates has a season-to-season preferred diet consisting of particular grass. Thus each of the twenty or more grazing species within an area utilizes the savanna in a different way from the others, and the total effect is complementary. The wild animals also wander seasonally and in this way spread their grazing and browsing activities over wide areas. Unlike domestic livestock they utilize the water supply without destroying water points through heavy trampling followed by erosion.

The selective feeding and rational use of the savanna by wild ungulates result in an ecological balance with the vegetation. It is even proper to speak, in a broad sense, of a giant symbiosis between the savanna vegetation and the savanna herbivores. A partnership with the savanna exists, and a maintained productive climax is achieved.

This state of balance is upset when the wild animals are replaced by livestock. The latter are almost always too numerous, and their continued heavy grazing exceeds the carrying capacity of the grassland. The grazing pressure is concentrated on a few species of grass that the domesticated animals prefer. One grass species after another is eliminated from the area, setting up a chain reaction that causes the whole plant community to change. Unpalatable, stiff herbs without nutritional value often take over. In heavily grazed and trampled areas the reduction of vegetation cover is so great that the soil is laid bare. Erosion follows automatically. The deterioration spreads over large areas and often leads to changes so drastic that previously fertile savannas are rendered worthless. Thousands of semidesert and desert areas in Africa and Asia are the end products of the negative evolution, which has taken place in a few decades. They are man-made wastelands.

Virtually all open range pasturage of livestock on marginal African lands results in lowered production and destruction due to overgrazing and overtrampling. Range management for livestock production in the African tropics and subtropics is not an easy task. Many agriculturists and veterinarians believe the problem is simply one of reducing livestock numbers, but it is much more complex than that. As noted, cattle, goats, and sheep do not behave like wild ungulates. They keep closely together. Grazing, drinking, licking salt, walking, ruminating, or sleeping, they never spread out over the grasslands—a habit that simplifies the herdsman's job. They literally wear away the vegetation on some areas of the range before moving to another, topographically less favourable site. It may be questioned if man will ever be able to manage marginal African grassland efficiently with the present narrow range of domestic livestock species. One who doubts the feasibility is the former Chief Agriculturist of Kenya, Leslie Brown. Most ecologists familiar with the problem in Africa agree with him. There is a considerable consensus that man should rid himself of the cattle-goat-sheep mentality, with its premise that domestic herds are the only way to produce meat in the tropics. Wild animals do that more effectively. Some of them can be tamed. Why not, besides the utilization of wild herbivorous mammals, make more effort to domesticate species that have evolved in the environment, and so are well adapted to it? The natural tropical vegetation will always constitute the chief fodder for protein-producing animals. Therefore it must be preserved, and the best way to do so is to preserve the wild herbivorous animals that are maintaining the savannas.

Overexploitation and Industrialization

As this book has repeatedly stressed, failure to undertake ecological surveys of the bioclimates, vegetation, and major habitat factors building up an ecosystem may have dire consequences for the exploitation and development of a region. Ecological ground plans may help to avoid mistakes, but it is essential to arrive at realistic interpretations of the surveys before making decisions to exploit land areas and undertake industrialization projects.

The industrial revolution has made increasing prosperity possible for many people and nations—and perhaps soon this will be the case over all of the inhabited world—but this trend has been accompanied by a deterioration of the environment in which people are obliged to live. Industrialization and overexploitation of natural resources have gone so far on every continent that they actually are impeding progress. Killing off vegetation over enormous areas destined for urban or industrial sites is ultimately a threat to human physical and mental health and to a host of other values. But there are solutions to these problems. For example, it is possible to spare much of living nature that is now destroyed simply because it is easier and cheaper to bulldoze an entire area than selected parts. Even if more costly, it is generally good economy for society to leave as many green patches as possible within developed areas. These islands of vegetation are not only a means of beautifying an exploited region; they may also serve as a water resource, a clean air refuge, an animal sanctuary, a site for human stimulation and relaxation, or a relict of the former charm of the landscape. All this may sound futile to an industrialist, but with the population increase and growing exploitation such remaining vegetation oases will be extremely valuable.

Vegetation Uniformity and Pesticides

An important aspect of grasslands and forests—in fact, all habitats with vegetation—is the danger created by simplifying the living environment. Cultivation automatically reduces the complexity of a habitat with its wealth of plant and animal organisms. Simplification is even the goal of agriculture, which is usually based on single or a few crop species. In animal husbandry and forestry, too, it is considered most efficient to favour growth of single or several particular species of grass for grazing or trees for timber products.

In general, the opinion that monocultures are the best form of agriculture, and locally also of forestry, is well founded when seen in a short-term perspective. They provide greater yields of particular crops and easier harvests produced more rapidly than do diversified cultures. But when regarded from the long-term perspective, the advantages may decrease; however, the evidence for this conclusion may take many years or decades, for forests even longer than that, to accumulate to where it becomes apparent. The slow or rapid deterioration, measured by the productivity of the crop concerned, is often avoided by such artificial means as chemical fertilizers, spraying of toxic chemicals, and other unnatural methods. Chemical pest and disease control, as well as advanced technology and genetic knowledge, has greatly improved crop production in monocultures, but new serious problems follow closely in their wake. These interventions in the environment, using artificial materials and means, have to be continuously, quantitatively, and qualitatively (in degrees of toxicity) increased to counterbalance diseases and pests. Biotic simplification and chemical manipulation over vast areas cause the artificial habitats to lose their former species diversity, which has been a guarantee of biological stability. Diseases and pests then invade the croplands and forests on an acceleratingly destructive scale, whereas in the past the habitat could cope with them on its own without losing productivity.

In modern times it is not just the monocultures themselves, covering thousands of acres of croplands and cultivated forests, that simplify environments and so create ecological hazards. The surrounding lands also are often drastically altered to suit one or another purpose of man. Adjacent vegetation is removed by mechanical or chemical cleaning for roads, buildings, fences, and so on. The biotic simplification comes to characterize entire regions, both above and below the ground. Man loses many biological allies that would have helped him to combat crop enemies without chemical warfare and to maintain a degree of ecological stability in the transformed ecosystem.

Artificial simplification is unavoidable in agriculture, though not necessarily in forestry. Obviously man must cultivate, sow, plant, and harvest, and with these activities he impoverishes the landscape and increases its ecological imbalance. The question is one of how far, in the long run, it is economic to go in transforming living, diversified environments to monocultures. Further, what is an acceptable price to pay for having to resort to repetitive toxic chemical treatments in place of lost biological stability? Ecologically the modern pattern of agriculture is an abuse of land, and places man in a biological unknown. Agriculturists claim that the monocultures and extensive use of chemical poisons are economically necessary. This may be so for the moment, but nevertheless the conditions threaten the future of mankind.

However, compromises are in sight that to biologists at least seem worth trying. The negative biological effects of monocultures could be considerably reduced if they were somewhat reduced in size so that they could be surrounded by natural or at least diversified habitats. A diversity of agricultural areas would benefit the farmer, other living organisms, the soil, and the waters —the whole environment, crops included—by producing plants that could reduce the pressure of parasites and diseases on crops and animals that are potential predators to many parasites. And man and other living beings would be less exposed to poisoning by toxic chemicals.

It is noteworthy that in modern management of rangelands the trend has changed from the old system of favouring a single species of grass to an increasing frequency of mixed plantings. A mixture of annual and perennial grass on the pastures extends the grazing period for livestock and provides increased productivity and ecological stability.

In forestry, the management philosophy has already come full circle. In North America and Europe forest monocultures were much favoured until recently. Deciduous trees were eliminated by chemical spraying over large areas. In Europe this system of maintaining conifer monocultures has been linked to outbreaks of harmful insects and other pests. Noxious insects and fungi have good opportunities to multiply and spread from one host to another when trees of the same species grow in close proximity. As controlling predators reproduce more slowly than their prey and are more vulnerable to pesticide treatments, they do not become effective controllers until the insects and fungi have done their damage. Moreover, in conifer monocultures the soil formation is adversely affected in the long run and the soil fauna is greatly impoverished, resulting in lower productivity of the whole habitat. The most modern practice is to diversify the forest stands by mixing hardwood trees with conifers. This favours stabilization of the fauna and reduces the disturbances from insect outbreaks. The situation comes close to that prevailing in natural forests. Thus the pendulum of forest management opinion has swung to the ecologists' side.

Elimination of the use of persistent toxic pesticides should be the goal. This was a key recommendation of President Kennedy's Scientific Advisory Committee on Use of Pesticides in 1963. Although much has taken place since then in many countries, leading to legislative and practical ameliorations, the pesticides are still in use over large areas. The problem they present is urgent. Governments, governmental agencies, agricultural organizations, and pesticide manufacturers have yet fully to meet their responsibilities in an all-out effort to deal with the biocide problems created by pesticide poisons.

In the United States about seventy thousand tons of DDT are spread per year, chiefly in agriculture. Toxic agricultural chemicals that by voluntary agreements or bans are no longer on the market in Great Britain and other countries because they have been proved to have more harmful than good environmental effects, are instead introduced and advertised on the markets of other countries—for example, in Africa and Asia. Pesticide manufacturers appear to be under no moral suasion to desist from poisoning and damaging environments as long as it makes profit for a few.

In this writer's opinion, the best treatment and discussion so far given this complex topic is by Robert L. Rudd in his book *Pesticides and the Living Landscape* (1964). One of his conclusions is: "Traditional pest control must yield to a new outlook, one that does not treat pest species as somehow independent of the natural forces influencing all living things. Viewed in this light, the conservation of one species and the protection against undue numbers of another are two sides of the same coin. The management of population numbers, for whatever purpose, requires close knowledge of the ecological factors that determine population density. We must therefore adopt the view that pest control is no more than the conscious lowering of numbers of undesirable species and that the word 'pest' itself has no meaning when populations are small enough to cause man no important inconvenience. Biological manipulations are much surer means to the end of achieving control of pests than are chemical applications."

Introduction of Exotic Species

Man has introduced exotic plants on every continent throughout history. Most important crop plants utilized by humanity are exotics in the sense that they are not native in most parts of their present range of harvest. In the United States nearly 200,000 named species and varieties of plants have been officially introduced. However, despite the enormous interchange of plants from one part of the world to another through accidental and deliberate introductions of exotics, a surprisingly small number of exotic species have actually established themselves in the wild. In the plant world, exotic introductions have often been successful, but there have also been many examples to the contrary. Exotics can be rigourously controlled in cultivated habitats, but as soon as they reach natural habitats by various means of dispersal, effective control is almost impossible.

In principle, all introductions of exotic plant species are a danger. They can bring quite unexpected repercussions and economic losses to man. No exotic plant should be introduced into a new environment without a thorough ecological investigation of both the area of origin and the area of proposed introduction. Exotic plants may react and behave quite differently in a new environment than in their home habitat. They may compete so successfully with native plants that they may displace or even exterminate the latter. They may upset balance among animal species by favouring particular species, and these may themselves become pests, cancelling out all the planned benefits to be derived from the introduction. It is generally better to try first with native species and give them a fair chance to propagate before bringing in exotics, which may modify habitats and ecosystems entirely.

Exotic plants may bring harmful organisms to a new area and thus indirectly cause much trouble, even catastrophes. Damage from diseases and pests of forest trees is greatest in cultures or plantations of exotic trees.

An example from Argentina shows what can happen when an exotic plant conquers a new habitat. The Canada thistle (*Cirsium arvense*) is despite its name a native European plant, but it has become thoroughly established in America. During the nineteenth century it invaded the pampas of Argentina. Within a short time the herds of cattle were starving, and the economy of the farmers was threatened. The thistle plants, freed from the ecological control of their original environment, grew higher than a horse's head and became so thick that cattle could not make their way through them.

In Australia, a wide spectrum of exotic plants has been introduced. Many of these species have dispersed and competed very successfully with native species. They have occupied areas that were denuded by detrimental farming and pastoral activities, making it impossible for native plants to recolonize the same areas.

On islands, the flora of which is often of evolutionary significance, introductions of exotic plants often exterminate or displace the native species. Tristan da Cunha, for example, has eighty-two alien flowering plant species, greatly exceeding the thirty-two or so indigenous species.

The climax vegetation of Mauritius is forest. Before man settled there, most of the island was covered by dense evergreen forests forming a closed canopy and descending from the mountain peaks to the sea. In areas with less rainfall there were woodland savannas intersected with marshes. In less than three hundred years this diverse and productive ecosystem of dominating forests has changed to a land under intensive cultivation, where the "natural forests" occupy 1.30 percent of Mauritius' total acreage. Yet these so-called natural forests consist to a great extent of exotic elements which compete successfully with indigenous trees and other plants. Hence the persistent expansion of exotic plant species threatens to eliminate the native ones before man has had opportunity to study them and judge their environmental and chemical values, which might be economically high. For example, in the past massive ebony trees of at least three species were a characteristic part of the Mauritius forests. Obviously these native trees were of great value, which was the very reason why ebony trees and other hardwoods were the first to come under the axe.

This tremendous environmental alteration in so short a time must have repercussions on the major components of the ecosystem. The local climate has probably undergone certain changes due to the removal of the forest cover. The hydrography of the island must have been affected. The soil formation has been drastically changed. Erosion must have increased. The disappearance of the natural vegetation, together with other man-made actions, has caused the indigenous fauna of Mauritius to fade out due to loss of convenient habitats. Of the island's twenty endemic birds, more than half are already exterminated. Of the remaining nine species, which desperately fight for survival in the few remaining habitats, all species except one are disfavoured by the inroads of exotic plants. Mauritius has recently lost three reptile species—of which one was the giant land tortoise—and five other species are at present approaching the same fate.

Exotic aquatic plant species in several instances have proliferated explosively with extremely rapid dispersal along waterways, both upstream and downstream. The most spectacular case is that of the American water hyacinth (*Eichhornia crassipes*), a species native to tropical South America. A few individuals of this species were found floating in the Congo River early in the 1950's. A few years later it had dispersed upstream—probably via boats—and downstream along the Congo and its tributaries. It literally blocked the boat traffic on rivers and lakes, as well as fishing activities. It changed the water ecosystem. Despite the expenditure of a million dollars in 1956–7 for control measures in the Stanley Pool, through which the Congo River flows, 150 tons per hour of *Eichhornia* were still passing Kinshasa on the Congo River in 1959. It took almost ten years for man to bring it under control in the Congo River, where it still occurs. This plant, usually introduced for ornamental purposes, has likewise spread with extreme rapidity elsewhere in Africa as well as in Asia and Australia.

Another South American plant, the free-floating fern *Salvinia auriculata,* has caused great trouble in Lake Kariba in Rhodesia and Zambia. The plant was found on the Zambesi River above the lake in the 1950's. By 1962 it occupied four hundred square miles of Lake Kariba. In Kenya's Lake Naivasha large areas have recently been choked up by *Salvinia,* threatening fisheries and recreational activities. In Ceylon this aquatic fern was received at Colombo University in 1939 and became naturalized nearby in 1942. Floods in 1947 spread the species so widely that twenty-two thousand acres of arable land and at least two thousand acres of waterways were infected.

The horn nut (*Trapa natans*), a European aquatic plant, was introduced into the United States in 1884. Although retreating in western Europe, it found optimal conditions in some New World watercourses, where it caused great difficulties. Some twenty years and an expenditure of several hundred thousand dollars were required to get rid of the horn nut in the Potomac River. The North American water pest *Elodea canadensis* has had similar negative effects in Europe.

Several other examples of introductions of aquatic plant species have seriously modified the habitats they invaded, and their eradication has cost tremendous sums.

Management and Landscape Planning

Management of vegetation and other natural resources is a conservation tool for areas that are used in a variety of ways and for different purposes. In this sense management is a kind of landscape planning, for every kind of land use modifies the landscape. Management may be applied to both natural and cultivated landscapes. In this writer's opinion, landscape planning should deal only with cultivated landscapes unless natural areas are doomed to such heavy exploitation as to lose their character. In that event, landscape planners may try to make the best of a tricky situation. However, they should always work closely with ecologists. Landscape architects have too often shown that they do not understand the biological values and qualities of a living landscape. They seem to think that they are modeling dead material. This is a dangerous concept. Sound landscape planning changes as little of the landscape as possible. The task, rather, is to protect it from modification, whatever may be the degree of cultivation, urbanization, or industrialization.

Every country should have a comprehensive system of landscape planning for both the countryside and the town site. The system ought preferably to be run by conservation authorities in close cooperation with economic and technical agencies. Although all planning aspects and requirements should be dealt with, the basic considerations and evaluations must be ecological. If they are not, the implications of and the natural forces within a living landscape will not be fully comprehended.

CHAPTER SEVEN

The Animals

Animals of our time are the culmination of millions of years of evolutionary development. They are distributed all over the world, in the sea and on the continents, including the polar caps. In both the northern and southern hemispheres, the number of plant and animal species decreases with increasing latitude. Consequently, the arctic and antarctic biota lack much of the diversity and complexity found in the tropics. Between these two biotic extremes lies a wide range of biomes, of which each is divided into a spectrum of habitats and ecological niches. This spectrum is ecologically almost infinite in variation due to numerous factors, among which the activities of man have an enormous environmental impact.

Wild animals are an integral part of all ecosystems of the earth. Although animals as a category evolved later than plants, they are at present of equal importance. Plants and animals, including man, are evolving together and are dependent on one another. The evolutionary success of flowering plants is in large part due to their close relationship with the animals. Successful reproduction and seed dispersal often require the assistance of insects, birds, and mammals. On the other hand, the animals cannot exist without vegetation. Both groups are pillars for the existence and function of ecosystems.

Man can use wild animal resources for his benefit in many ways. They are highly efficient and economically significant food producers. Many species also yield hides and skins. A great number of species contribute to the maintenance of productive habitats and ecosystems. All animals are of major scientific significance.

Many play a vital role in human recreation, and so have a social value. More and more people like to see, watch, study, and photograph animals. Others enjoy hunting and fishing. The concept that wild animals are an essential part of a living landscape and of the cultural heritage of nations is spreading among people all over the world.

Ecology

Most animals are environmentally sensitive because of a high degree of adaptation and specialization that makes them vulnerable to habitat changes. But when such alterations occur, there are always other species favoured by the change.

The ecology of animals is intrinsically complex, but there are some fundamental and simple rules relating to interspecific and intraspecific competition. In the case of the former, no two species can occupy the same niche. If two species having the same food and shelter needs—to put their ecological requirements simply—come together in the same environment, they will automatically compete. But even species having the same ecology are differently equipped to cope with the innumerable components of the environment—climate and predation, for example. Therefore the species best adapted to the existing environmental conditions will be successful and will survive. By pressure the weaker species will be eliminated from the habitat. Often it will adapt to another, less favourable niche that still provides basic needs. There are numerous instances in the animal world where such an ecological hierarchy exists, with the ecologically "strongest" species occupying the most favourable niche. If that species is eliminated from that niche, a formerly suppressed species will promptly occupy it. An example of such a competitive relationship between species is given by the trout (*Salmo trutta*) and the char (*Salvelinus alpinus*). Where one of these species is solitary, it spreads and occupies various habitats, but wherever it is joined by the other, competition ensues. The two species then confine themselves to limited ranges offering varying conditions for each. In higher altitudes the char is favoured by its arctic adaptation. It seems to compete successfully with the trout in running waters in arctic and subarctic areas, but seldom or never in lower regions, where the trout dominates (Curry-Lindahl, 1956).

In the relatively simple communities prevailing in Iceland the ptarmigan (*Lagopus mutus*), which is the only grouse occurring on the island, has a much wider habitat and food range than in similar areas where the near related willow grouse (*L. lagopus lagopus*) or red grouse (*L. l. scoticus*) is also present.

In savanna habitats of tropical Africa the ecological hierarchy among twenty to thirty species of grazing and browsing herbivorous mammals is even more pronounced.

All animals are directly or indirectly dependent upon green plants, which through complicated processes provide carbohydrates, proteins, and fats to herbivores. The animals that convert plants to meat serve as food for predators. But not all predators consume herbivorous animals. Many species feed on predators, forming additional links in the food chain.

The food chain concept seems simple, but the nutritional requirements of vertebrate animals are quite complex. The basic need is fuel for energy. This is provided chiefly by carbohydrates, which the herbivores break down through digestion and metabolize to calories of energy. Fats, proteins, calcium, phosphorus, amino acids, and other nutrients are also essential in a proper balance to maintain metabolic health.

There is a relationship between body size and energy turnover as well as between body weight and the rate of protein turnover. The maintenance requirements of protein per unit body of weight increases with decreasing body size in the same way the requirements for energy increase. This means that in ruminant herbivores with comparable digestive systems, the maintenance of body weight by smaller animals requires an intake of diets with a higher proportion of protein and soluble carbohydrates (Bell, 1970).

The daily calorie requirements of vertebrate species depends on their size, physiology, and biological year cycle. Man needs about two thousand to three thousand calories per day, while the nutrient requirements for growth and antler development in the white-tailed deer is six thousand to ten thousand calories per day. Warm-blooded animals like mammals and birds require more nutrients than do cold-blooded ones like reptiles, amphibians, fish, and invertebrates, because most species in the former group utilize a high proportion of calories to maintain a constant body temperature. Moreover, the nutritional requirements of an animal may differ greatly at various seasons of the year. They are also modified directly or indirectly by climate and the presence of other animals such as symbionts, parasites, and pathogens.

Herbivores and omnivores have the most complex nutritional metabolism. Since carnivores feed on meat, a well-prepared and balanced diet that contains most of their basic nutritive needs, their metabolic processes are much simpler.

Only a few aspects of the ecology of animals have been briefly touched upon in this section. A fair understanding of this must necessarily be based on knowledge in detail and depth of how species are adapted to their respective environments. To learn this requires continuous research. At the present time man is just beginning to know and understand a little of the ecology of the animals around him.

Habitats

Wild animals can survive only if suitable environments for their life requirements are preserved. In ecology and zoology the term habitat, or biotope, means the type of environment in which an organism lives. The fact that a species prefers a particular habitat is generally an indication that the area represents optimum or at least favourable environmental conditions for that particular animal. On the other hand, few species encounter in one habitat the optimum conditions for all of their functions.

Usually a host of animal species, or populations, live in the same habitat. They are all interrelated in various ways and form communities. Similarly, all these animal species are related to plants, and to nonliving organisms, within the habitat. Every part of the environment that exerts an influence upon the life of a species belongs to the habitat of that species. Moreover, each species functions as a part of the conversion system of its own habitat. This complexity explains why it is virtually impossible to foresee in detail what will happen when a habitat is exposed to artificial disturbances or alterations.

Under natural conditions the successional development in habitats takes its own course, often leading to a climax situation. In areas where man's influence is dominant, an unnatural succession takes place, often leading to destruction of a highly organized biocommunity. Too many examples throughout the world illustrate how habitats and ecosystems break down due to human factors, such as elimination of wild animals, introduction of exotic species, use of fire, agricultural methods, overgrazing and trampling by livestock. As discussed in the previous chapter, habitats influenced by one or several of these unnatural factors may degenerate, particularly in the tropics and subtropics, through soil erosion to a state of sterility.

In literature on wildlife management the habitat may be referred to as the "cover" in which an animal exists. In this sense cover commonly refers to the physical complex of water, soil, and vegetation in which an animal functions. This is an oversimplification because, as explained above, the term habitat must necessarily include all biotic and abiotic features of the environment concerned. Climate, air, and other animal species are also important components of the habitat of a species.

This means that the ecological niche is a part of the habitat, and the habitat is a part of the ecosystem. While hundreds of animal species may utilize the same habitat in a broad sense, only one species can utilize an ecological niche. The habitat can be defined easily by a description of its physical character, while the ecological niche can be referred to properly only by describing the character and the function of the animal species that makes use of it. Features like a species' behaviour, structural adaptation, physiology, nutrition, feeding technique, interspecific relations, and so on provide information that tells more about the ecological niche of a given species than does an attempt to describe the niche itself. However, the distinguishing features of a niche are often too elusive for human perception.

A species may shift habitat seasonally or with increasing age. Some species are so extremely specialized to a habitat that their existence depends absolutely on maintenance of the habitat, while other species are flexible and adapt readily to changed conditions within the habitat. A species may also occupy different types of habitats in various parts of its distribution range. Since many habitats and animal populations are dynamic, new ecological adaptations are constantly evolving in many species due to competitive pressures or successful expansion in range.

Although it has long been recognized that the ecosystem plays an essential role in the evolution of species, ecologists seem to have paid little attention to the role of long-term evolutionary processes at species level as working dynamics of ecosystems and habitats.

A thorough knowledge of the functioning of ecological niches and habitats is necessary to understand the role of individual species and how they operate within an ecosystem. It is also essential to know the tolerance limits of each species, e.g., physiological response to temperature, humidity, light, water composition, and so forth, as well as interrelations with other living organisms. A considerable number of limiting factors, physical and biological, determine the ecological range of a species during successive life stages—eggs, embryos, larvae, juveniles, adults. The processes of evolution, habitat selection, and adaptative radiation have led to a maximum efficiency for survival in each species. The limiting environmental factors not only control the species and its activities but also its abundance and distribution.

Extreme specialization to a habitat may be dangerous and has in several cases led to extermination. In fact, the most successful species are those that are not particularly specialized. Man is the best example. Among other mammals the elephant, buffalo, leopard, red fox, and field vole show high capacity to settle in widely different habitats. Among the birds, a number of migratory species are ecologically very flexible. Many species shift twice a year between arctic and tropical habitats. En route between these two extremes the birds utilize temperate and subtropical habitats.

Home Ranges and Territories

Home ranges and territories can be individual, shared by a family, or collective. For many species the home range is equivalent to the territory. Many species defend their territories against members of their own species in various ways. The function of territorialism in animals is to avoid overcrowding and overutilization of the habitat by the same species. It also serves as a means of dispersal. This spacing mechanism is an important factor in an understanding of the interactions between animals and their environment.

There are different opinions of the role of territorial behaviour (dominance and spacing behaviour) as a limiting and regulating factor to breeding populations. Most workers (Errington, 1956; Tinbergen, 1957, 1958; Wynne-Edwards, 1962; Watson and Moss, 1970) think that territorial behaviour is such a limiting factor, while Lack (1954, 1966), working chiefly on birds, believes it is not. Finally, Hinde (1956), also basing himself on birds, could not find any unequivocal field evidence that it does limit numbers, though he thought the evidence strongly suggestive of this.

An excellent example of such a limiting mechanism is shown by an Australian gecko, *Gehyra variegata*. In this lizard, behavioural regulation (through aggressiveness) of female numbers per homesite (independent of food supply), in conjunction with male territoriality, determines population size (Bustard, 1970). In this case the mechanism of regulation of population size results in a carrying capacity below that at which food would become limiting. This prevents food shortage and results in a more stable population size than would be the case if the population was food limited.

Population densities of many other animal species are certainly held far below the starvation level by similar behavioural-territorial mechanisms.

In the zoological sense, a territory does not necessarily mean a range that is defended by an individual or family or a group of families (in this book called a clan) against other individuals of the same species. For some species a territory may be collective and used as feeding area by a population that is settled elsewhere individually or in families or in groups living in different nest sites, if such are used. Within their home sites individuals and families of rodents belonging to the same clan may exhibit "microterritorial" tendencies by defense and aggressiveness during their stay in the site burrow or its immediate vicinity. This behaviour varies in different species and is also probably influenced by the physiological state of the individuals, as shown in many species of vertebrates.

Burt (1943) defines his concept of home range and territory in mammals as follows: Home range "is the area, usually around a home site, over which the animal travels in search of food. Territory is the protected part of the home range, be it the entire home range or only the nest."

There are also "mini-territories" that are defended in connection with displays during mating times. Such mini-territories are known in all classes of vertebrates. Territory-holding is fairly common also among invertebrates.

Home ranges and territories of a given species change in size and shape due to various factors—topography, microclimate, access to cover, nest sites, feeding areas, and last but not least, population density in relation to food supply. Furthermore, the territory is not always fixed within a definite area, even during the breeding season. The territory may become elastic and may occasionally be replaced by a new area. In general, however, a restricted territory is usually to be expected during the breeding period.

As mentioned in the section on habitats, many migratory species shift their centers of activity from one season to another.

They therefore have two or more different home ranges. For some marine mammals, many birds, and several fish, these various home ranges may be separated by thousands of miles.

Productivity and Biomass

The preceding sections have briefly discussed the ecological setting of animals—their ecological niches, habitats, home ranges and territories, as well as how these units function. This ecological background is of importance for understanding the productivity of animals. Essential factors in measuring the productivity of a wild species include reproduction, nutrition, growth rates, age of sexual maturity, liveweight gains, social relations, such general ecology items as water requirements, movements, and migration, and meat and hide yield.

Generally, animal species seem to reproduce as fast as they are able. Presumably their numbers are primarily limited by density-dependent mortality rate factors such as food shortage, predation, and disease. The reproductive rate has evolved, through natural selection, in relation to the availability of food for the breeding adults and other environmental factors. In birds, which feed their young, the reproductive limit is clearly marked by the clutch-size which likewise has evolved through natural selection in relation to the maximum number of young for which the parents can provide food without detriment to themselves (Lack, 1954, 1968). However, populations may not respond directly to their food supply just through breeding or mortality rates. Other factors such as social structure of populations and related behavioural mechanisms may also interact in this process. For example, when an animal dies of starvation it may be due to the territorial behaviour of its congeners, which have excluded it from a feeding area. In this case behaviour is the limiting factor and starvation the proximate factor.

The reproduction rate of wild mammals is high and is adjusted in a rational way to what the environment may support. In most areas it is far superior to the productivity of livestock. The key factor in determining population density of animals and their biomass is for most species the available food supply within the habitat.

The productivity of wild animals may be measured in several ways. At a given time of the year any habitat contains a total quantity of living material of both plants and animals. This totality is called the standing crop. In terms of weight it is referred to as the biomass. The animal biomass is usually given as the weight of live animal matter per acre. In general, herbivorous animals dominate in the biomass of vertebrates within terrestrial communities. The standing crop biomass of mammals living within a given area is calculated by multiplying the number of individuals of each species by the minimum weight of the female adult.

The total biomass of, say, a dense forest is extremely difficult to measure due to the complexity of the habitat and the number of plant and animal species involved. Open grassland communities are easier to measure.

The standing crop varies from one habitat or ecosystem to another. In general, the highest standing crop of terrestrial and freshwater habitats is to be found in tropical areas, the lowest in arctic regions. In marine habitats, the situation is different. Arctic or antarctic seas are seasonally very productive and in this respect may exceed subtropical and tropical waters. Marine productivity may have a bearing on shoreland communities of vertebrates.

The animal biomass of an area is not necessarily proportional to the plant biomass. Although a moist tropical forest represents a high plant biomass and also a high number of animal species, it often has a lower animal biomass per unit of area than does a semiarid savanna. At present, the highest known animal biomass in terrestrial habitats is produced by tropical savannas in Africa. In North America, the natural plain grasslands of the past produced the highest animal biomass ever found on that continent.

The energy conversion cycle and rate of metabolism of various organisms within a habitat or ecosystem are important elements when measuring the productivity per unit of area. These transfers of energy are in their turn connected to the rate at which organisms are produced and consumed. Here two main categories of organisms, plants and animals, are working within the habitat or ecosystem, contributing to the productivity of the area in different ways. The rate at which plants produce new organic material through photosynthesis and other chemical processes is called the primary productivity, while the rate at which animals convert plant organisms to animal tissues is called the secondary productivity.

A low biomass of organisms having a high rate of metabolism and rapid reproduction may support a much higher biomass of consumers (either herbivorous or predatory animals) having a low turnover rate and long life expectancy. Such a situation leads to a high secondary productivity.

The size of the standing crop within a community also influences the efficiency of production. In some habitats there may be too few organisms to utilize the energy available. The situation can also be the reverse—an inadequate supply of energy available to each organism. This is often the case in lakes, as indicated by stunted fish populations. Hence a large standing crop is not always the same as a high total productivity. An interesting example of aquatic productivity influenced by fish predation was given by Hayne and Ball in 1956. When fish were present in a pond, production of bottom fauna on which they fed was very high. In the absence of fish, the production rate of the bottom fauna decreased and finally stopped, with a larger standing crop than when the fish were present. Thus the standing crop of bottom fauna was depressed in the presence of fish, but production increased. In ponds with fish, the annual production amounted to 811 pounds of fish food and 181 pounds of fish per acre. This represents an efficiency of 18 percent in energy conversion from bottom fauna to fish (Smith, 1966).

Productivity, standing crop, and biomass are bases for measuring the efficiency of habitats and ecosystems. They provide tools to compare the animal productivity of natural areas with that of areas influenced by human activities, or of entirely man-made lands. They give important economic data indicating which is the best form of land use on a sustained yield basis. Obviously, such comparisons of animal productivity are not significant when made in areas or regions that are not ecologically comparable. The most useful data on biomass are those obtained from studies of long enough duration to include fluctuations in climate and populations.

The data in Table 1 of year-long standing crops (biomass) from various parts and habitats of Africa give an idea of magnitude and differences. Only the larger species utilizing the habitats analyzed have been included. The real figures of productivity are higher. Most of the data from Table 1 have been obtained in national parks and nature reserves. The two top biomass figures as well as Nos. 4 and 5 come from two contiguous national parks: the Queen Elizabeth National Park in Uganda and the Albert National Park in the Congo. The tables show clearly that wild African ungulates utilize poor natural grazing lands in such an effective way that they reach a biomass far superior to that of cattle on similar lands, and almost as high as cattle in excellent, heavily fertilized artificial pastures. Moreover, the wild ungulates constantly produce a high yield of meat without destroying the soil and vegetation upon which they subsist, while cattle in similar habitats destroy the environment by overgrazing and trampling.

The reasons for this difference, which is pronounced throughout the tropics and subtropics and especially in Africa, is that wild grazing and browsing ungulates make use of almost all edible vegetation of the habitat. Elephants, hippopotamuses, rhinoceroses, buffalos, giraffes, the array of antelope species from the giant eland to tiny dik-diks, hogs, and so on feed on virtually all available plants and eat almost all parts of the vegetation: roots, grass in various stages, seeds, shoots, leaves, thorny bushes, bark, foliage of medium-sized trees, and so on. In short, the wild animals utilize almost the whole biomass produced by the habitat. They interplay with the annual rhythm of the vegetation, which is not overutilized. Movements and seasonal migrations spread the animals and their grazing pressure over the ecosystem, and this shifting grazing maintains the system's immense productivity. The situation can be compared to a kind of giant symbiosis between vegetation and animals.

Domestic livestock behave quite differently. They depend on only one or a few plant species, which they devour until these plants are virtually exterminated in the habitat. Then they turn to the next palatable species and the process of elimination is repeated. Cattle graze in concentrated troops and use the same water holes day after day. The adverse effects of trampling on the vegetation and soil combined with the grazing almost always leads to serious erosion and decline of the habitat. Several other factors that help to explain the great difference of productivity between wild animals and domestic livestock on marginal lands will be discussed further in the two sections that follow.

The great variety of faunal composition in African grazing ecosystems may possibly also be explained by substitutions in the grazing succession among herbivorous animals in relation to minor differences of vegetation structure. Such a differentiation of the diet selectivity in ten to twenty species of ungulates sharing a habitat has great flexibility and could be a basis for the high efficiency of pasture utilization of tropical savanna ecosystems.

The association between grazing ungulates is so close that it is not easy for observers to find to which degree one or another species is dependent on the presence of other species. In the Serengeti National Park in Tanzania herds of wildebeest and

TABLE ONE

Standing Crop Biomass of Wild Ungulates on Grasslands Ungrazed by Livestock

Area	*Number of species*	*Range type*	*Biomass per sq. km. in kg.*	*Author*
Uganda	7	Grass and bush savanna (partly overgrazed)	34,944	Petrides & Swank, 1958
Congo	10	Grass and bush savanna	24,406	Bourlière & Verschuren, 1959
Tanzania	7	*Acacia* woodland	21,937	Watson & Turner, 1965
Uganda	11	Grass and bush savanna	18,795	Bere, 1960
Congo	1 (hippopotamus)	Grass savanna	16,800	Curry-Lindahl, 1955
Kenya	Several	*Acacia* savanna	15,760	Talbot et al., 1962
Uganda	9	Grass and bush savanna	13,360	Petrides & Swank, 1958
Kenya	18	Grass and bush savanna	13,215	Ellis, 1961
Tanzania	Several	*Acacia* savanna	6,550–12,310	Lamprey, 1959
Uganda	7	Grass and bush savanna (undergrazed)	11,100	Petrides & Swank, 1968
Kenya	9	Grass and bush savanna	8,257	Petrides, 1956
Congo	5	Forest-savanna mosaic	5,950	Pirlot, 1956
Tanzania	15	Grass and *Acacia-Commiphora* savanna	5,283	Grzimek, 1960
Kenya-Tanzania	Several	*Acacia-Commiphora* bushland	5,250	Talbot, 1963
Rhodesia	15	Mopane woodland savanna	4,418	Dassmann & Mossman, 1961
Kenya	9	Grass and bush savanna	2,180	Petrides, 1956

TABLE TWO

Standing Crop Biomass of Livestock on Natural Grazing Lands

Area	*Range type*	*Biomass per sq. km. in kg.*	*Author*
Argentina	Pampas	14,000	Leplae, 1933
United States	Prairie	11,000	Leplae, 1933
South Africa	*Themeda* high veld	8,500	Leplae, 1933
Kenya	*Themeda* savanna	3,500–5,500	Henderson, 1950
Rhodesia	Savanna	5,000	Leplae, 1933
Congo	Savanna	4,000–5,000	Leplae, 1933
Kenya	*Acacia* savanna	1,960–2,800	Talbot, 1963
Kenya-Tanzania	*Acacia-Commiphora* bushland	370–1,350	Talbot, 1963

TABLE THREE

Standing Crop Biomass of Livestock on Artificial, Fertilized Grazing Lands

Area	*Range type*	*Biomass per sq. km. in kg.*	*Author*
Belgium	Cultivated meadow	45,000	Leplae, 1933
Rwanda	Montane cultivated land	40,000	INEAC *
Congo	Rain forest	34,000	INEAC
Belgium	Cultivated meadow	14,000	Leplae, 1933
Congo	Cultivated savanna	3,600	INEAC
United States	"Excellent rangeland"	5,500	U. S. Soil Conservation Service

* Institut National des Etudes Agronomiques au Congo Belge

zebra are closely associated. It has been found there that the wildebeest is favoured by the grazing activities of the zebra. The latter eats tall, coarse, and cellulose-rich grass, which facilitates the grazing of the wildebeest. Thomson's gazelle (*Gazella thomsoni*) frequents only areas where the herb layer is kept short through grazing by other ungulates (or through burning). The ecologically very flexible buffalo (*Syncerus caffer*) usually selects leaves from larger species of grass, which the antelopes in general do not use. In some areas the buffalo are of great importance to smaller herbivores by breaking down the very tall stands of *Echinochloa pyramidalis*.

Most wild ungulates feed selectively like livestock, but the diversity of the wild species, their sizes, and different feeding techniques make them complementary to each other as grazers and browsers in tropical habitats, and they do not cause environmental damage. But even in temperate biomes, where the number of wild ungulate species is less than in the tropics, there is no damage to the vegetation caused by selective feeding. Deer or various species are capable of selecting the most nutritious plants not only of certain species but also among plants of the same species or portions of the same plant. Where food becomes limited, deer can reduce and temporarily even eliminate locally preferred plant species of high forage quality, which causes the animals to move to areas where similar plant species occur.

All examples in Table 1 come from tropical Africa, for which the most data is available. It is evident that in the present time undisturbed savannas in Africa produce the world's highest biomass of terrestrial wild animal populations. What is the situation in other parts of our globe? The only comparable areas are the South African velds and the North American prairies of two hundred years ago, where the wild ungulates certainly reached a biomass of 35,000 kilograms per square kilometer (approximately 15 short tons per square mile)—the maximum biomass in Africa—and presumably exceeded it. The savannas of tropical Africa undoubtedly supported much larger populations of ungulates one hundred to two hundred years ago than they do today.

The uniqueness of the tremendous productivity of African savannas becomes clear from a comparison with the present biomass in other parts of the world. A forest-meadow mosaic in India, Kanha National Park, heavily grazed by domestic livestock consisting of elephants, cattle, and buffalo, also supports nine wild ungulate species. The latter reach a biomass of 938 to 1,178 kilograms per square kilometer (Schaller, 1967). The number of wild ungulates has been depressed by the presence of too many livestock individuals, which exceed the wild animals both in number and biomass. The combined biomass is 3,880–4,100 kilograms per square kilometer, at the price of a rapid decline in the quality of the habitat. In order to restore productivity to the level of its potential, it would be necessary to remove the livestock from the area. The available data indicate that the Kanha National Park is capable of maintaining a permanent biomass of about 3,700 kilograms per square kilometer of a variety of wild ungulates under optimum conditions, after the exclusion of cattle and domestic buffalo from the habitat (Schaller, 1967).

In the United States the National Bison Range in Wyoming, a prairie habitat, supports a biomass of 3,400 kilograms per square kilometer. This figure is certainly inferior to the biomass of the former prairie, which in addition to herbivores—bison, pronghorn, and deer—included millions of prairie dogs (*Cynomys*).

Caribous of the Canadian tundras have a biomass of 800 kilograms per square kilometer. The deer forests in Scotland yield a biomass of red deer on the order of 1,000. An almost primeval montane forest in Czechoslovakia supports a biomass of herbivorous mammals of about 500. The great herds of the saiga antelope represent the optimum usage of the Eurasian steppes of the southern U.S.S.R. Yet the biomass is only about 350 kilograms per square kilometer during the winter period, when the highest concentrations of saiga occur, despite a population size varying between one and a half and two million.

Up to the present the protein productivity of wild and domestic ungulate populations has been studied more thoroughly in Africa than elsewhere. Although Africa, favoured by its wealth of highly productive ungulate species, reaches unequalled biomass figures, it is likely that in other parts of the world where the habitat is not destroyed, the potential, natural sustained yield productivity of marginal lands is far superior to that of livestock or agriculture on the same lands despite the cost of heavy investment and loss of capital through habitat decline. Much more research on natural productivity and "wild protein" is needed. Investigations in alpine, arctic, and high level regions of the northern hemisphere have shown the high forage quality of tundra and forest floor vegetation, which in turn accounts for the rapid growth rates of the ungulates such as reindeer, moose, and other deer of these northern haunts.

If other terrestrial animals that in many parts of the world are used as human food—e.g., rodents, birds, reptiles, amphibians, and invertebrates—are included in the biomass, the figures become overwhelmingly superior to those for livestock. However, few data exist for vertebrates other than the larger mammals. The fact that on the steppes around the River Volga in southern U.S.S.R. there may be 32,500 susliks (ground squirrels) and voles (meadow mice) per square kilometer as against 0.4 saiga antelope, which have a biomass of 350 kilograms per square kilometer, indicates the importance of small mammals in the biomass of some areas. The biomass of eleven species of amphibians in a montane rain forest in Rwanda in tropical Africa was 2,100 (Curry-Lindahl and Lamotte, 1964), a minimum figure. Invertebrates may also represent high biomass values.

A savanna of *Andropogon* grass in Guinea gives a biomass of 25,000 (Lamotte, 1947), and montane savannas of Mount Nimba in the same part of Africa yield 18,700 kilograms per square kilometer (Roy, 1952). Some examples of vegetation biomass, expressed in dry weights (kg. per sq. km.), are given below for comparison:

Primeval lowland rain forest, Congo	100,360,000 kg.
Secondary lowland rain forest, Ghana	36,236,000 kg.
Deciduous 120-year-old forest of oaks and beech, Europe	27,500,000 kg.
Secondary lowland rain forest, Congo	25,000,000 kg.
Mangrove forest of *Rhizophora mangle,* Puerto Rico	6,300,000 kg.
Grassland savanna (herbs only), Tanzania	1,400,000 kg.
Prairie of *Andropogon* grass, United States	655,000 kg.
Marsh of sedge *(Carex),* United States	465,000 kg.

The figures above and those in Tables 1–3 together show the tremendous disproportion between the biomasses of chlorophyll vegetation and the herbivores that consume this type of food.

Another important aspect to bear in mind when examining the animal productivity of various areas is the rate of turnover of ungulate populations—the metabolic conversion. A high biomass does not necessarily mean a high turnover or an extremely high dividend in the form of meat. When high biomass values are represented by ungulate populations with a high proportion of rather slow-growing species that reach sexual maturity at a rather late stage and have a long gestation period, such as the elephant and the rhinoceros, the turnover rate may be less than it is in an area with a lower biomass composed of fast-growing, early-maturing, and fast-reproducing ungulates, such as the buffalo, antelope, and hog.

These facts should not, however, be taken to mean that the slow-growing mammals ought to be eliminated and fast-growing ones substituted for them. The latter cannot fill the same ecological niches and do not influence the environment in the same way as the slow-growing, large species. It is the diversity of species that makes the habitat or the ecosystem so effective as a productive unit.

The hippopotamus utilizes lake, swamp, and river margins and moreover plays a vital role within the ecosystem by bridging the conversion cycle between aquatic and terrestrial habitats. The elephant is one of the most important factors in the ecology of various habitats. It is a landscape maker contributing greatly to age-old changes of vegetation. It opens up pathways in dense forests by breaking down trees so that leaves, fruits, and bark become available to other grazers, browsers, and seed-eaters, and it aerates the soil by digging up underground water during drought periods, making it available to numerous other species from mammals to insects. Its droppings spread tree seed over wide areas. The black rhinoceros feeds extensively on young gall acacia, preventing the savanna from becoming sterile bush.

Environment Carrying Capacity

Each area of grassland or forest has a carrying capacity beyond which it cannot be utilized by animals or man without causing damage, deterioration, and decreased productivity. In undisturbed areas inhabited by natural populations, various mechanisms regulate the animal numbers in relation to the environment so that they remain within the habitat's carrying capacity. In areas altered by man and used by him or his domestic livestock, overexploitation often results, producing and causing the habitat to deteriorate.

The simplest definition of carrying capacity is the maximum number of animals of various species that can make indefinite use of the habitat or the ecosystem without damaging it. It is chiefly a question of interrelationships between vegetation and the animals, but obviously other factors, such as climate, water regime, soils, and predators, are also involved. The carrying capacity is not static. It is as dynamic as the habitat or the ecosystem itself.

Most range damage anywhere on the globe—tundras, prairies, pampas, steppes, and savannas—is due to man's ignorance or neglect of the carrying capacity. Overutilization leads to destruction. Lush grasslands are changed to eroded areas, where deserts are encroaching on formerly productive lands, wiping out useful animals and interfering with the hydrographic cycle.

When the carrying capacity of various rangelands is studied in the light of thousands of human failures to plan, use, and manage them properly, one may question whether livestock has any place at all on arid rangelands. With few exceptions the available data and experience seem to indicate that in the long run cattle, goats, and sheep are uneconomic in arid and semiarid regions, inevitably producing such serious damage to vegetation and soil as to ruin the landscape. Where wild animal populations are still allowed to utilize comparable arid lands, animals and landscape both flourish, producing a sustained high yield. Even though figures on carrying capacities from various tropical areas of the world have to be taken as approximations, they definitely indicate that "wild habitats" used by wild mammals, reaching tremendous biomass values, remain within the carrying capacity of the land. Just the opposite situation prevails for most tropical land utilized by domestic livestock. Despite the fact that the biomass of livestock is considerably lower than that of wild mammals, the domestic animals "appear in virtually every case to exceed the carrying capacity of the land" (Talbot, 1966).

Therefore, whether it involves wildlife management or animal husbandry, or both, it is essential to adjust animal populations to the carrying capacity of the habitat or the pasture. Without doing so, man cannot expect to restore the formerly productive areas he so effectively has destroyed.

Wildlife as a Source of Food

Man uses wildlife as food over much larger areas of the world than is generally realized. The tropics in particular have vast regions where wild animals provide the only available protein crop. Moreover, the potential of wild animals as a food resource is enormous. Preceding sections and Tables 1–3 have shown that savanna lands support a biomass of wild herbivorous animals up to fifteen times higher than that of domestic animals.

A classic example of a livestock population crash due to mismanagement of habitats and overstocking comes from Kenya. In the

Kajiado District in Masailand the total number of cattle was estimated at 347,000 in 1942–44, 450,000 in 1955, and 680,000 in 1960. It rose to a peak in 1961, but in that year, due to overgrazing, drought, and floods, the Masai lost 65 percent of their cattle, the number falling to an estimated 203,000. The government was forced to spend large sums on famine relief in Masailand. This catastrophe was the direct result of disease control leading to overpopulation of livestock. Artificial water supplies were developed without commensurate improvement in rangeland management through the control of stock numbers and movement (Brown, 1968). In contrast to the hundreds of thousands of dying cattle in the misused and devastated lands of the Kajiado, the wild animals —zebras, antelopes, elephants—were in a remarkably good condition due to their extraordinary physical resistance to harsh environmental situations.

The importance of animal resources was emphasized at the UNESCO Intergovernmental Conference of Experts on the Scientific Basis for Rational Use and Conservation of the Resources of the Biosphere held in Paris in 1968. The conference stressed that the problems of wildlife management deserve high priority in scientific and economic planning for maintenance of an ecological balance favourable to human well-being. It was noted that the need to make efficient use of animal protein resources has never been so urgent as it is today, and that the global shortage is serious. Further, the conference pointed out that while animals form an essential part of the ecosystem in many efficient forms of land use, most of the sixteen major domestic animals are not chosen for their ecological suitability but to satisfy traditional demand for well-known products, and out of familiarity with basic techniques of animal husbandry. Wildlife may well contribute as either a supplementary or major form of land use—primarily as a source of protein, but also of leather, fur, recreation, and attraction for tourists. There has been considerable recent progress in arriving at an understanding of how a wide spectrum of wild herbivores utilizes the environment rationally and produces large quantities of protein and other products from degraded or marginal lands. Ecological knowledge of wild animals as a source of food should be fully used in planning development projects.

At the Biosphere Conference attention was drawn particularly to the great array of ecologically adapted African herbivores that function in maintaining stable ecosystems and producing a high biomass, with more than thirty species coexisting in some areas. In most cases these are marginal lands with poor soils and vegetation cover. Since optimum conditions for most domestic livestock are narrow, environmental conditions adverse to their production prevail over large parts of the world. Nevertheless numerous and repeated attempts have been made to introduce or improve livestock production on ecologically unsuitable lands, at great cost and with very little economic return. Artificial environmental changes made to meet the requirements of domestic livestock have usually overlooked the inability of such animals to adapt to adverse climatic conditions. Moreover, the detrimental effects of these changes on the land in reality mean loss of productivity and destruction of habitats. Using animals that were well adapted to the original conditions would have been economically more rational. Such animals almost always exist in the area involved. They are the wild animals that have evolved as products of the very environment that man tries to alter by introducing domestic species exotic to the ecosystem.

Previous sections on productivity, biomass, and carrying capacity have given data showing that on many rangelands, particularly in tropical countries, wild animals make more efficient use of the environment than do domestic livestock, and that it is possible to maintain high production on a sustained yield basis without adversely affecting the habitats and the ecosystem. The protein potential of wild animals on vast areas of the global surface is much higher than that of livestock. Several facts explain this. By now it is apparent that the prime factor is the marvelous adaptations of wild animals to local conditions through effective interplay with the environment. Diet, nutritional efficiency, and water requirements of the animals are parts of this adaptation.

Another factor, the reproduction rate, is determined by the age at which animals reach sexual maturity, by the length of the gestation period, by the frequency of births, by the litter size, and by the age at which animals normally cease to reproduce. These vary for different species, but for most wild animals the reproduction rate is remarkably high. Moreover, for many species the rate is governed by the environment. In years when the food is abundant, the reproduction rate goes up. In bad years it goes down. This reproductional flexibility is of great environmental importance and contributes to balance in the ecosystem.

Most species of African wild ungulates—even larger species like the buffalo and the eland, largest of the antelopes—normally produce at least one young a year. Many species begin to breed when under one year old. This, for example, is the case for all the gazelles (more than ten species), impala, and numerous other species of antelope that reach an adult weight of less than about 200 pounds. Ungulates with adult weights ranging from about 250 to over 1,000 pounds breed when just over a year old, the females producing their first calf at two years. Species that do this include the eland, which emphasizes the tremendous protein value of this ecologically adaptable species, and, for example, topi, kongoni, and wildebeest (*Gorgon taurinus*).

The buffalo breeds when 2 to 3 years old, the hippopotamus and the black rhinoceros at 6 to 7, and the African elephant at 12 to 15 years. For cattle the average breeding age is about 3½ years, but many native-owned cattle in Africa do not breed until they are about 5. Under exceptionally favourable conditions with artificial means, maturity age may be reached at about 2½ years. Sheep and goats may start breeding at about age one.

Gestation periods of larger wild ungulates such as eland, roan, and waterbuck do not differ from that of cattle—nine months. However, a number of large and all the medium-sized antelopes have shorter gestation periods than that. It is remarkable that such a large species as the hippopotamus has a shorter gestation period than cattle. Gestation periods of some African herbivores are given in Table 4.

The length of the gestation period is less significant for the reproduction rate than the frequency of giving birth during the life span of a species. Almost all wild herbivores with gestation periods less than a year produce one calf a year. That is, for instance, the case with such large species as the buffalo and the eland. Some species may even have twins. For the bush pig five is a normal litter, but any number up to ten is common. Moreover, this species breeds twice a year. The warthog may do the same, but usually has only two to four hoglets in a litter. Several of the smaller antelopes also breed twice a year.

Many wild ungulates breed throughout the year and are able to restore their reduced numbers rapidly. In a reserve in South Africa, for example, the number of springbok (*Antidorcas marsupialis*) increased from 247 to 698 during the period from July, 1966, to April, 1968, despite some cropping (Van Zyl, 1968).

In another part of South Africa, an impala ewe was observed in the wild during the course of her entire lifetime. She presumably died at the age of 14 years. She had lambed even in drought years and had produced thirteen lambs (Skinner, 1969). In the very

TABLE FOUR
Gestation Periods of Some African Ungulates

	Months		Months
Duikers (several species)	4	Scimitar oryx	8
Dik-diks (several species)	4	Hippopotamus	8
Bush pig	4	White-tailed wildebeest	8½
Dorcas gazelle	5	Wildebeest	9
Redfronted gazelle	6	Waterbuck	9
Springbok	6	Eland	9
Warthog	6	Roan	9
Dama gazelle	7	Buffalo	11
Bushbuck	7	Burchell's zebra	11–12
Greater kudu	7–8	Somali wild ass	12
Sitatunga	7½	Black rhinoceros	12
Hartebeest	7½	Nubian wild ass	13
Blesbok	7½	Giraffe	14
Reedbuck	8	African elephant	22

large herds of wildebeest in the Serengeti-Mara region of Tanzania and Kenya more than 95 percent of all adult females in the population produce a calf each year (Talbot, 1963).

The mortality rate in wild herbivores is generally regulated by predation and other environmental pressure. Although the longevity of a considerable number of species is known, few data are available showing at which age wild animals cease to reproduce. The life span may give an indication that the reproductive capacity in certain species continues up to relatively advanced ages. Among African ungulates the elephant may reach an age of about 60 years, the rhinoceros and the hippopotamus over 40, the giraffe 28, the oryx 18, the buffalo 17, the waterbuck 16½ and the wildebeest 11. However, few individuals reach their potential maximum age, because as soon as they begin to decline physically, they succumb to predation and/or disease. This is a positive mechanism in maintaining a healthy population.

Comparative growth rates of wild ungulates and domestic livestock from Africa clearly indicate that the wild animals are also superior in this respect. Most species of ungulates reach maximum weights and marketable, i.e., economically harvestable, size at earlier ages than domestic animals. Cattle under native management in East Africa reach marketing size in 5 to 7 years, but under more efficient management might do so in about 4 years. Thomson's gazelles require roughly 15 months, Grant's gazelles and impalas about 18 months, topis and kongonis about 2 years, wildebeests 2½ to 3 years, and elands about 3 years (Talbot, 1966). Even so large a species as the buffalo grows more rapidly than cattle. Sheep and goats require 1½ to 2½ years to attain marketable size.

The great difference in growth rates between wild and domestic ungulates grazing or browsing in similar habitats is explained by a higher assimilation of food among wild herbivores. A striking example of this phenomenon is given by the eland, which can attain a weight of 665 pounds in two years on pastures where domestic cattle would literally starve to death (Riney, 1960). A weight of 665 pounds is superior to the average adult weight of cattle.

Anatomical features show that wild ungulates have a greater nutritional efficiency. This is indicated by the lesser development of their digestive tract, compared with domestic ungulates of comparable weight.

The average liveweight gain reflects the nutritional efficiency of wild ungulates in comparison with domestic ones. For the eland it is, for instance, more than twice as much—0.73 pounds per day—as in cattle—0.30 pounds per day. In other large-sized species, such as wildebeest, kongoni, and topi, it is much higher than in cattle and other livestock (Talbot et al., 1962). Again, the hippopotamus needs much smaller quantities of food—23.6 calories per pound of body weight—than cattle to build up its enormous body mass of about one ton (von Anghi, 1940).

The meat yield—the proportion of meat obtained per body—from various species of wild herbivores in comparison with cattle is another important factor to consider in analyzing the productivity of a rangeland. It is expressed in percentage in relation to the live weight of an animal. In African-owned cattle, it seldom exceeds 50 percent and is usually below 40 percent. In zebu cattle in Kenya it is 32 percent. Long-legged tropical meat goats average about 45 percent and sheep about 44 percent (Williamson and Payne, 1959). In wild animals the average meat yield is much higher. In hippopotamus it is 70.9 percent, in Grant's gazelle 63.2 percent, in impala 61 percent, in kob 59.1 percent, in eland 59 percent, in Thomson's gazelle 56.8 percent, in buffalo 55.2 percent, in springbok 55 percent, in topi 53.6 percent, in kongoni 52.5 percent, in wildebeest 50.6 percent and in zebra 50 percent (Haarthorn, 1961, Ledger et al., 1961, Mann, 1962, Talbot et al., 1962, Van Zyl, 1968).

Moreover, the meat yield in carcasses of wild animals contains far less fat than those of domestic animals. At a meat yield of 51 percent, wild animals have a fat content of 1.8 percent, while cattle have 14.1 percent. For a meat yield of 60 percent, wild animals contain 0.3 percent fat and cattle 28.4 percent. In the springbok, for example, the meat yield in males may exceed 62 percent, but the fat is less than 1 percent. The production of lean meat protein is a more efficient version of fodder than is the production of fat, which is the most uneconomical of animal products. Thus the wild herbivores, with virtually no fat, make much better use of the vegetation than do cattle. Under comparable environmental conditions, wild animals provide an excess of protein over that obtained from cattle equivalent to 6,000 calories for every 100 pounds of liveweight. These data show that wild ungulates are far more efficient in utilizing the available nutrients within a habitat than are cattle.

The diet of the grazing and browsing wild animals helps to explain why so many species can live together, utilizing the same habitat with no damage to the land, whereas the same habitats cannot tolerate cattle, goats, and sheep.

The water requirements of wild herbivores differ from one species to another, but almost all of them have a remarkable ability to withstand drought. Most species drink when water is available, but they can also be active for days without drinking. Several species—among them oryx, Dorcas gazelle, Grant's gazelle, Thomson's gazelle, gerenuk, and impala—spend months in areas where no surface water is available. Other species of ungulates apparently never, or only exceptionally, drink water. They seem to be able to extract the free water in green plants in sufficient quantities for subsistence even during prolonged dry seasons or in desert areas with extremely low average rainfall and very high temperature. Several desert antelopes are examples of herbivores that seldom or never drink water. Some desert rodents, such as the kangaroo rat, for example, manage to survive on water obtained through their own body metabolism. This adaptation of ungulates to conditions where water is scarce or even lacking makes them particularly well-suited for arid and semiarid regions. Whereas shortness of water supply is a tremendous barrier to domestic animals, it does not limit the populations of a number of wild ungulates. These can make full, long-term use of the habitat to their own advantage and that of man as well.

The resistance of many wild animals to extreme drought conditions is but one aspect of their general environmental resistance capacity. Others are their natural resistance to starvation and to parasites and diseases, environmental factors to which domestic livestock are highly vulnerable. During extreme drought years it is spectacular to see how healthy the herds of zebras and antelopes in Africa are and the larger marsupials in Australia, compared with livestock frequenting the same habitats. The latter are literally starving to death, and locally entire herds may be wiped out.

Wild animals have been used by human beings as a protein resource for thousands of years, and the utilization of wildlife as a renewable natural resource continues to be economically justified. In most marginal and submarginal lands of tropical and subtropical regions it is even the best form of land use. African and Australian experience has shown convincingly that rationally planned cropping of wildlife is feasible, producing high sustained yields of both meat and hides. The technicalities underlying game harvesting operations vary from one area to another. The cropping technique depends on the species involved, the climate and topography of the country, and the distance to the nearest settlements. As in all cropping schemes, processing, food hygiene, storing, and marketing, as well as sociological aspects, are important factors that have to be considered.

Well-organized, economic game cropping has been carried out in a number of African countries on an industrial, long-term scale. The demand has been so heavy that most of the meat has been absorbed by local markets in the vicinity of the cropping area—as a rule within about one hundred miles. Various large mammals have been utilized: in Uganda hippopotamus, elephant, and kob; in Kenya elephant, wildebeest, and zebra; in Zambia elephant and red lechwe; in Rhodesia impala, kudu, wildebeest, duiker, zebra, buffalo, and warthog; and in South Africa (Transvaal) impala, wildebeest, kudu, springbok, blesbuck, duiker, bushbuck, zebra, and warthog.

Only a few African countries have thus far initiated large cropping schemes, and relatively few species have been utilized. The potential productivity of African ungulates is so high that cropping targets can be increased many times when refrigerated transport facilities and other technical facilities become available.

In Australia wild animal cropping schemes have been economically successful. In recent years the country has exported weekly fifty-five tons of meat from three species of kangaroos. But in this case it appears that the exploiters have drawn so heavily on the animal populations as to deplete them. This is probably intentional, because marsupials are accused of competing with sheep.

Cropping of wildlife can also be carried out in the form of game ranching, and in some areas game and livestock ranging can be combined economically. In South Africa and Rhodesia game farming is now a common practice. Many ranches produce higher quantities of first-class meat at lower cost after they have shifted from livestock to wild animal production, and the net profits to the land owners increase considerably. The costs of maintenance and harvesting are negligible. Some ranches in Rhodesia base their production on sixteen or more species of wild ungulates—one reason for their success.

Ranchers have found that soil and vegetation grazed by wild animals have recovered from previous misuse. In Transvaal a rancher working severely depleted land discovered that whereas 8,800 pounds of cattle per square mile had degraded his land, 17,600 pounds of springboks and bushbucks on the same area enabled the land to recover and produced more profit (Riney, 1963). In Transvaal about 3,000 game ranchers were operating successfully early in the 1960's, many of them on land previously depleted by livestock. In other parts of Africa the potential for such game farming is even greater.

Cropping and game farming have produced both fresh and dry meat that has sold readily. Canned game meat of antelopes and hippopotamus has even been exported overseas and has been accepted in Europe and America. However, it is essentially to meet the home protein need of African countries that the local game meat is so invaluable on that continent.

Game meat is quite suitable for human use. Its quality and flavour are in no way inferior to beef from cattle provided it receives the same treatment. In Africa the meat of all ungulate species as well as many other wild animals is consumed and enjoyed by both Africans and Europeans. The author has eaten fresh meat from more than fifteen species of African ungulates and found them all excellent. He considers the flavour of meat from several species superior to that of domestic livestock.

In most African countries the wild animals are the only, or the principal, supply of animal proteins. Despite the depleted wild animal populations, the consumption of bush meat is remarkably high. In Liberia it is estimated that between 60 and 70 percent of the present animal protein production is based on game animals; in some regions the figure is 100 percent. In Ghana over 80 percent of the animal protein consumed is in the form of wildlife. In the Congo (Kinshasa) are vast regions where about 60 percent of the human population base its nourishment on bush meat. In the Congo (Brazzaville) 50 percent of the fresh meat consumed is produced by game animals. In many parts of East Africa meat from wild animals is likewise by far the most important source of protein, though precise figures are not available. Dry bush meat (biltong) is for sale in the markets of almost every village and town in tropical Africa. In the cities such meat usually fetches higher prices than other forms of meat, probably because bush meat is scarcer but more appreciated than other meat. When fresh bush meat is marketed, it is promptly sold, indicating how esteemed it is.

The foregoing shows that it is not unrealistic to envisage a rational exploitation of wild animal populations in Africa for protein production. Economically and ecologically, this is the best form of land use on marginal and submarginal lands—that is in areas unsuitable for agriculture and pastoralism due to climate and soil. About 90 percent of African savannas can be classified as marginal lands that when left intact, despite their soil poverty and the low rainfall, are extremely productive and yield more meat than rangelands with the highest livestock-carrying capacity. This emphasizes the great economic importance of wild animals as a source of protein. In addition, the value of their hides is high. It is tragic for Africa, and for mankind in general, that lands that could be so productive are rapidly destroyed by inappropriate use.

The greatest single danger to Africa's wild animals comes from the herds of ecologically unadaptable, tick-ridden cattle competing for grazing space on semiarid pastures. The cattle threaten the animals by threatening the land itself, and ultimately they threaten man. It is a dramatic battle of time. The destruction goes on. Will governments become convinced in time that a high production of proteins can be constantly maintained from wild mammals on lands that deteriorate under other forms of use?

While Africa is exceptionally well stocked with highly productive wild herbivores, other areas of the world also can produce wild animals as an additional source of wealth. Many parts of Asia,

Australia, South America, as well as the temperate regions of North America and Europe, can be used more rationally than at present.

The saiga antelope of the dry steppes and semideserts of the U.S.S.R. is a spectacular example of how a species natural to an area can prove a source of revenue. After they were almost exterminated in the 1920's, sound management raised the population to about 2 million. About one-quarter of these live between the Volga and the Ural rivers, where the population is about four times as dense as in Asia. In this region alone professional hunters kill between 150,000 and 350,000 saigas annually, representing about 6,000 tons of excellent meat, more than 2,000,000 square feet of hides; edible and inedible fat as well as horns. Despite this killing, the number of saigas is increasing. In view of the sparse vegetation of the dry steppes, this productivity is astonishingly high. Production of saiga antelopes has proved to be the best form of land use in these arid regions. In the same way, the prairies of North America, the arid plains of Australia, the semideserts, savannas, and forests of Asia, and many regions of Europe could profitably produce wildlife. With a new conservation policy and wise management that included cooperation with wild animals, many areas could certainly reach a higher annual productivity and economic yield than they achieve under present conventional forms of land use.

Wild birds too can be farmed. The best example comes from Iceland, where eider farms capitalize on the demand for eiderdown, and eggs are also harvested. The birds are quite wild, but they are not disturbed by the cropping of down and eggs. This is due to a long and close relationship between the breeding eiders and man. A careful "eider husbandry" has not upset this relationship. The eiderdown production is based on numerous managed colonies of eiders, where the utilization of the birds' products is the privilege of the landowner.

Recreational Value of Wildlife

The discussion in the previous section has dealt chiefly with the role of wild animals as a protein resource. But wildlife proves to be of high economic value in other ways as well. Tourism and recreation in many countries are largely dependent on wildlife. This is notably the case in Africa. For many years tourism ranked as Kenya's second highest industry, exceeded only by coffee export. Then in 1968 Kenya's tourism income figure surpassed that of the coffee export value, to become the largest source of national revenue ($44 million in foreign exchange). In 1970 the foreign exchange earnings from tourism (232,000 visitors) in Kenya had risen to $53,249,700 (an increase of 11 percent over 1969). It is evident that the majority of these overseas tourists come to Kenya to visit the national parks, the nature reserves and the beaches of the Indian Ocean, where there are two marine national parks. In Uganda tourism is the fourth biggest industry, in Tanzania the sixth. In the 1970's it is estimated to be East Africa's biggest industry, grossing about $75 million in Kenya alone.

The well-organized tourist industries of East Africa and South Africa have shown convincingly the enormous economic value of wildlife. While other parts of the continent have the same potential, many areas are still virtually dormant in the field of tourism development. The eastern Congo has an extraordinary variety of spectacular landscapes, habitats, and flora and fauna that represent a high touristic potential. Likewise the animals of West Africa have great possibilities as tourist attractions.

Economically, Africa cannot afford to jeopardize the future of so tremendous a capital resource as its wild animals and their habitats. If properly managed, these renewable resources will remain longer than the copper belts of Zambia and Katanga, the gold in South Africa, and the diamond fields in the Congo and South Africa. And what has been said of Africa in this respect can be repeated for all other continents.

It is not only in Africa that the economic potential of wildlife as a recreational resource is well exemplified. Many examples could be given from other areas of the world. In the United States the recreation based on wild animals for watching, hunting, and fishing is the basis for multimillion-dollar industries as well as for federal and state budgets. For example, 2,020,885 deer (mainly white-tailed deer) were shot by licensed hunters in the United States in 1966. The license fees not only paid the state wildlife department expenses, but they also brought in a great deal of money as revenue.

Predation

The two preceding sections, dealing with the nutritional and recreational values of wildlife for man, were so placed because these aspects relate intimately to the ecological factors of productivity, biomass, and carrying capacity discussed in the sections that went before. The discussion now returns to the ecological setting of animal populations.

Predation is one of the most important factors in the structure of animal populations sharing the same habitat or ecosystem. It may look simple: one animal kills another. But it is not so simple to judge the effects of predation on populations. What is well understood is that predators are a very necessary part of the natural world. They have existed for 400 million years. Yet a balance has been maintained between predators and prey. So far as is known, no species has by predation depleted another until man appeared on the scene of life.

Few issues in wildlife management are more controversial than the role of predation and the relationships between predatory and prey populations. Hunters and farmers exaggerate the predation pressure on animal populations they themselves find useful or enjoyable. For centuries they have tried to eliminate predators by shooting, trapping, poisoning, bounties, and so forth. On the other hand, conservationists have been too generous in crediting predators with always being a beneficial factor in keeping down destructive animal populations.

Still, many people think that if they shoot predators the latter will become scarce in the area concerned and the prey animals they want to favour will become more abundant. Conservationists have for decades claimed that destruction of avian and mammalian predators will not result in an increase of game animals that man wishes to kill himself.

Predation is a dynamic tool that functions quite differently in various situations. Usually a predatory species exerts no or little numerical influence on the prey population. The predator chiefly strikes at surplus animals, which in any event would have been eliminated from the habitat through intraspecific competition and other ecological factors.

To all appearances nature produces a superabundance of life forms. But the mortality is great, and in reality only a minority of individuals of a species reaches maturity compared with the number born or hatched. Through a selective process developing over thousands of years, the reproduction of every animal has adapted in the way that best serves the survival of the species, but not necessarily of the individual. The reproduction surplus of a species thereby benefits many other species and constitutes an important factor for maintaining equilibrium in an animal community.

It is a general rule that the predatory animal is inferior in numbers to the categories on which it preys. If the prey species—voles or mice, for example—has an extremely high population density, the predatory animals—foxes, weasels, birds of prey, owls, and others—can respond to this abundance with larger litters than usual. This process in turn contributes to reestablishment of the prey group's lower, more normal population. However, much evidence favours the view that the peak of the population curve of small rodents is sooner or later lowered even without the effect of predatory animals, although more slowly (Curry-Lindahl, 1959, 1962).

Practically all investigations into the effect of predators on natural populations of their prey animals show that neither in the long or short run does predation have any effect on the mean population size. It is therefore usually meaningless for man, by bounties and propaganda, to encourage the decimation of carnivorous mammals and birds with the object of affording greater scope to the prey animals that he may want to utilize.

It is true that predators can temporarily and locally press down populations of their prey species, but it seldom if ever occurs, except when man comes into the picture, that predation leads to total extermination. In cases where predators specialize on a certain prey species or group of species, the predation frequency usually becomes synchronized with the supply of the prey animal. Apparently, however, the mortality percentage of the prey species through predation is no higher when its frequency curve is at a peak than when it is at low level. During high fertility periods the predation strikes in the first place the superfluous individuals—the ones that have difficulty in finding shelter and food in the intraspecific competition.

Recent field and laboratory work suggests that as the density of prey increases, the numbers taken by a predator rise to a plateau level beyond which further increase in prey density has no effect on the number of prey consumed (Holling, 1965; Murton, 1968).

Should the predation include prey species other than those on which the predator is specialized, as sometimes happens by chance, the effect on the population figures of the prey species will be slight or none. However, it does happen that when a predator's normal prey is rapidly reduced in number, or perhaps even temporarily disappears from the territory, the predator is forced to turn to another kind of prey. The choice can then fall on a species more vulnerable to predation than is the normal prey animal, and the effect may then be evaluated locally.

One still hears the nineteenth century idea with respect to predatory animals, that kill or be killed is nature's law. The struggle between different animal categories or between individuals of the same species is usually of an indirect nature. Kill *and* be killed is an expression that comes closer to reality, for the majority of animals must kill other living organisms—either plants or other animals—to be able to live themselves. Their struggle is to *find* food and to escape *becoming* food. A wild animal seldom dies of old age. Instead, the animals form links in a nutrition chain or a predation chain that is often topped off by man.

The nutrition chains, varying constantly, dependent upon seasonal and environmental conditions and upon the biology of the animals concerned, may be simple or complicated. As a general rule an equilibrium principle prevails, for each link in the chain is not only dependent upon the others but also constitutes an important building stone that if removed may cause the collapse of the whole pyramid. But nature soon builds up a new one, always with great quantities of various life forms as a broad base and frequently with a predatory animal species at the top.

The reproductive rhythm of animal species in the course of the year is often characterized by frequency peaks at the close of the reproductive period. Starvation, disease, predators, wanderings, accidents, and so forth, later gradually decimate the number of individuals to reasonable proportions that the vegetative environment can support. Broadly speaking, the populations are self-regulating.

A natural, limited geographic area can only harbour a certain amount of plants whose density, power of growth, and individuality are determined by the environment—i.e., space, soil, water, light, competition with other plants, and relations with the animals, whether positive or negative. It is the same with the animals. Thus an intimate relation prevails both among the animals and between animals and vegetation.

Since nearly all organisms produce more offspring than the home territory can harbour, or than are needed for the survival of the species, the surplus plays an important biological role. Otherwise the whole process would be meaningless, which nature never is. One may say that the importance of the surplus of a species is to serve as food for other species. It is accordingly the surplus that is decimated by the predatory animals. Where these are not present, starvation, disease, and other natural interferences account for disappearance of the surplus, limiting the remaining animals to a number in harmony with the environment. This milieu resistance is a fundamental rule met with everywhere in nature—from sea to lake, from plain to forest, and valid also for man.

The predatory animals are in the front line among the consuming or decimating environmental factors as controllers of the prey. They constitute the first state of preparedness, although not always the most important. Starvation, or rather malnutrition, comes later —in this case a synonym for interspecific as well as intraspecific competition. Diseases in a broader sense, as a rule, break down the population numbers first when the other two reducing factors are unable to absorb the surplus rapidly enough. Seldom will the surplus become so large that all the components of the milieu resistance are mobilized at the same time. As a rule the predators can master the situation. Malnutrition (starvation) paves the way for diseases (often parasites) which kill weakened individuals. When the same individuals were healthy and well-nourished they probably carried the same diseases, but at that stage they were unharmed. This seems to be the schematic picture of what happens in nature (cf. Curry-Lindahl, 1964), where the environment ultimately controls animal populations.

The environment of a predator, like every area populated by living organisms, is limited in its capacity to produce plants and animals, exactly in the same way as every farm is limited to its biological productivity. As noted, nature produces surplus animals that are unable to survive in the habitat where they are born. Some of them disperse and populate new areas. Through intraspecific competition other individuals die of starvation or become so weak that they cannot resist sickness and disease. Surplus populations are also reduced through predation, and this is probably what usually happens within a habitat in equilibrium. That predation primarily affects the surplus individuals of the red grouse (*Lagopus lagopus*) in northeastern Scotland is the finding of long-term investigations carried out by Jenkins, Watson, and Miller (1963). This research has clearly shown that predators have no significant effect upon the population fluctuations of the red grouse. In fact, winter predation on surplus non-territorial red grouse was seven times as heavy as it was on territory-owners. The latter are also much less affected by mortality from starvation and disease.

Other data obtained from population studies on other birds as well as mammals confirm that it is chiefly surplus animals that are fed upon by predators, but that this predation strikes various species at different times of the year, depending on reproduction cycles and seasonal territorial behaviour of the prey populations. Surplus individuals are in any case doomed to die at an early stage of their life.

Careful analysis of the food habits of fresh- and brackish-water fishes have shown that most species are opportunistic to some extent and that true regulators do exist in the sense that they will quickly dampen major outbreaks of prey species (Darnell, 1958, 1961, 1964, 1970).

To summarize the factors limiting higher vertebrate populations, and to simplify complex phenomena, food availability, competition (for food and biotopes), habitat selection, territory, and predation, in varying degrees, influence the dynamics and regulate the numbers of animal populations.

Predator-prey relationships belong to the trickiest problems in biology. There is no general explanation of how they function in detail, because the situation always changes from habitat to habitat and from time to time, depending on the constantly changing population structure within biocommunities.

Many studies of the food habits of raptorial birds in different habitats—wilderness areas as well as cultivated landscapes—have indicated convincingly that the common assumption, "the only good bird of prey is a dead one," is without foundation and completely misleading. This discriminatory attitude has often underlain legislative measures condemning all birds of prey and owls. The same attitude prevails toward carnivorous mammals. Such legislation is still in force in many parts of the world. During recent decades, however, a drastic change has occurred in legislation in this field. Almost every year the list of predatory mammals and birds protected in various countries is increased by one or several species. In 1969, twenty-two states in the United States had laws protecting all their birds of prey, and twenty-five states protected all but a few species.

Birds of prey may be victims of other predators, but in this respect no species is so important a controlling factor as man. Hence, man's concern with protective measures for birds of prey is a matter of dealing with his own activities.

If the concept that predators preying on game mainly take the surplus of prey populations is accepted, one may ask if the game productivity of an area from man's point of view would not be increased if natural predators were replaced by man. The answer in general must be no. Man is unable to prey in the same effectively selective way as an animal predator. On the contrary, he would act negatively on the species he is preying upon, lowering its genetic and constitutional quality, because lack of natural pressure will in the long run cause a deterioration of animal populations.

A recent but already classical example of the positive role of predation on population quality comes from the Isle Royale National Park in Lake Superior, where in a study of wolves and moose it was found that the moose population was very healthy and had twinning rates of 38 percent, but before the wolves were present on Isle Royal, the moose had overpopulated the island and the declining population had an estimated twinning rate of 2 percent (Murie, 1934; Mech, 1966). In Alaska, deer from populations controlled by wolves and heavy winter snows are larger than deer from comparable areas without wolves and with mild winters.

Legislative measures alone are not enough to preserve carnivores and birds of prey. Research, education, and propaganda are necessary so that all groups of people may understand the biological role of predators. For instance, it is important to avoid classifying these animals as "harmful" and "useful" species. Their ecology and function in biotic communities cannot be thoroughly understood without detailed studies of their necessities, the nutritional value of their different prey species in various habitats in all the seasons of the year, their hunting techniques, and their choices of prey as well as predator-prey relationships, population dynamics and cycles, and so on. At present not very much is known about these subjects, though a considerable amount of material has been accumulated in various countries, especially in the United States and Africa.

An important part of the predation problem concerns the choice of the individual victim. Does a raptor attack a particular animal or does it take its prey at random? Many records indicate that hunting predators tend to single out animals with physical defects. Such animals are more vulnerable and apparently induce attack, which often ends in death for them. This would mean that predatory species have a selective effect through their choice of prey. To determine whether such a selection occurs, it is necessary to see how the attack is made and in some cases also to be able to examine the dead victim. Although opportunities to make such observations do not occur often, so much evidence of selection of injured and abnormal prey has accumulated that it is unlikely the choice of weak prey always has been merely coincidental. But despite the evidence of selective predation by predators, much more research work must be carried out before the significance of this factor can be estimated objectively.

As noted, predator-prey relationships are only one of several factors regulating animal populations. However, it is clear that many authors dealing with the role of predation as a population depressant have underestimated its importance. As far as presently known, it is the only limiting factor that is functioning continuously, in all seasons, at various population densities, in different habitats, in wilderness areas as well as in highly cultivated lands. The degree of constant predation is in proportion to the density of the prey populations. Predation functions as a normal environmental factor, and since it is related to prey densities, the number of predators present within an area is determined by the same factor. Relatively few predators are adapted to subsist only on one species of prey, but even in such cases predation should not be regarded as a true limiting factor on the prey population. It is merely the other way around: the available prey regulates the population density of the predator. Thus from a biological viewpoint it is needless to control the predators.

The data obtained from thorough, long-term investigation on birds of prey and owls in Michigan and Wyoming in the United States by John and Frank Craighead (1956) have strongly indicated that predation in a greater or lesser degree plays an important role in establishing and maintaining a state of balance in animal populations. They conclude that predation by birds of prey is a biological process that tends to prevent excessive increases of prey species.

On the other hand, we know from the Serengeti plains in Tanzania, where the world's largest assemblages of big savanna mammals occur—sometimes in herds up to 400,000 ungulates—that the predation of lions, leopards, cheetahs, hyenas, wild dogs and jackals, and numerous other carnivores is negligible as a mortality factor in these ungulate populations, which live in harmony with the vegetation without destroying it through overgrazing or overbrowsing. The controlling factor here is diseases which normally are harmless but in certain periods of the year, when the nutritional values of grasses for certain species of animals are at their lowest, turn fatal and become the most important mortality factor for just these animals.

Although predator-prey relationships are still far from being fully understood, the implications may be summarized as follows: Predators have a valuable function in natural animal communities and do not normally reduce prey populations to low levels. Predation is an inhibiting factor on population explosions of species and locally can even keep down excess populations to the extent that explosions are prevented.

In well-balanced ecosystems predation contributes, together with other environmental factors, to maintenance of prey populations at levels that are in harmony with the capacity of the habitat. Predation is selective and functions as a biological control by eliminating diseased, wounded, or weak animals; it maintains quality, health, and high genetic standard in prey populations. It is a useful environmental tool that man can utilize in managing game populations. Most confusion concerning predators and their predation has come about as a result of the idea that economic and biological effects on a population are identical. It is hardly possible to evaluate predation economically without taking into consideration its ecological and biological mechanisms, effects, and significance. Predation must be recognized as a natural law.

Control Problems

Eradication campaigns against predatory animals and birds and other so-called pest animals, with bounty payments to encourage people to kill predators indiscriminately, have been carried out for the professed purpose of preventing damage. But ecological research has seldom investigated to determine if the species concerned really does damage. Where this has been done, the results have in almost every case been a rehabilitation of the species accused of being injurious. It is true that tigers, lions, leopards, jaguars, cougars, lynxes, wolves, coyotes, wild dogs, wolverines, and bears are capable of killing livestock. And foxes, jackals, raccoons, skunks, polecats, martens, badgers, otters, minks, and other medium-sized carnivores can kill game birds or sport fish. However, this does not mean that these predators are noxious as species. Their general ecological role in an ecosystem or habitat must be mapped out so that it can be judged in its full context. Moreover, one or a few categories of people—for example, hunters or sheep owners—should not be permitted to decide whether a predatory species is beneficial or noxious. Other interests in human society should also be consulted and respected.

It was not many years ago when the majority of predatory mammals and birds were not protected at all and bounties were paid for the killing of them. During the last ten to twenty years—the timetable is different in various countries—with increasing knowledge of the true role of predators, the attitude has changed.

The bounties are vanishing, and protection is now given to carnivores, birds of prey, and owls in many countries on all continents. But there are still countries which continue to hang on to the old system, because people by tradition believe predators are harmful and antagonistic to human interests, and accuse these animals of killing domestic stock or animals man himself wants to hunt or fish.

In South Africa, for example, several beneficial or indifferent carnivores are still legally persecuted despite the fact that if they have to be classified as useful or harmful animals they definitely count in the first category. A study (Bothma, 1971) on the food habits of some carnivorous mammals from southern Africa shows this clearly. For example, the Cape fox or silver fox (*Vulpes chama*) is annually killed by "jackal clubs," institutions and individuals, because it is claimed that this species is damaging domestic stock. In view of the diet of the species this policy and accusation seem to be highly questionable. Sixty-six Cape foxes were collected in areas of abundant sheep farming. Yet no evidence of domestic stock was found in any of the stomachs. The food items found in the stomachs of silver foxes were almost entirely rodents, plants (chiefly grass, wild fruit berries, and seed), and carrion. The Cape fox should obviously be protected in South Africa. Also, the side-striped jackal (*Canis adustus*) is often regarded as a threat to domestic animals. However, available scientific data indicate that this is not at all the case. Carrion and wild fruit are the two major components of this jackal's diet, which also includes reptiles, birds, and insects. Another predator, the bat-eared fox (*Otocyon megalotis*), feeds entirely on insects, carrion, rodents, reptiles, and amphibians. The civet (*Viverra civetta*), a large member of the carnivores, is chiefly a fruit eater, but feeds also on insects, carrion, birds, rodents, molluscs, spiders, and grass. Finally, the aardwolf (*Proteles cristata*) bases its economy almost entirely on insects and carrion. The latter is at present protected in Transvaal.

Livestock is rarely preyed upon by wild predators. This happens always in extraordinary situations, or where an individual has discovered that domestic animals are easy prey. Man's warfare should be declared against that individual and not against all individuals of that particular species.

Predator control on a large scale has had disastrous and undesirable effects in many cases. This can be illustrated by the now classic predator eradication scheme that took place in the Kaibab Forest of Arizona and began in 1907. In order to protect the decreasing population of mule deer within the forest, it was decided to exterminate wolves and cougars and to reduce the number of coyotes. The eradication operations were successful, but they resulted in an overpopulation of deer. Over the next seventeen years the deer population rose from about 4,000 to 100,000 animals. This led in turn to permanent damage to the forest vegetation and to the topsoil, which caused a food shortage for the deer and resulted in starvation, disease, and weakened condition that brought about a tremendous population crash. From 1924 to 1926 about 60,000 deer starved to death. Numbers continued to decline for a long time afterwards. The forest and soil were ruined for a long period ahead, and parts of the habitat were irreversibly damaged. In 1939 the deer were levelling off again at around 10,000 animals. This management error in the Kaibab Forest showed convincingly how useful the role of predators is in maintaining healthy animal populations in balance with the carrying capacity of the environment. Once the habitat equilibrium is tampered with, it may be impossible to cope with the process of deterioration.

A somewhat similar story affected the moose in Isle Royale National Park in Lake Superior after the extermination of wolves, but in this case the situation was reestablished when wolves recolonized the island and restored the population balance between vegetation, moose, and the wolves themselves (Mech, 1966).

Many similar examples are at hand. The local eradications of leopards in Africa have resulted in a pronounced increase of baboons, pigs, and hogs that ravage cultivated fields and destroy crops to such an extent that people want the leopard back.

Under normal circumstances most vertebrate animals are beneficial or useful to man, but occasionally the population density of a species rises to levels where damage of economic significance occurs. In some cases even public health is threatened. Among vertebrates, rapidly multiplying rodents are the animals chiefly regarded as pests. Some species of bats carry rabies and may therefore be fatal to man. Among birds some corvid and seed-eating birds are locally offensive to human interests. Snakes are often considered generally dangerous, and indiscriminately killed.

Before control measures against a "pest species" are applied, it is essential to investigate the local problem. Frequently there has been failure to ascertain what is happening within the biocommunity where the pest species occurs. In many cases the "culprit" species has not even been identified correctly, and control operations have been undertaken against a wrong species. Too often action has taken the form of indiscriminate destruction of all small rodents, or all bats, or all seed-eating birds of an area.

An ecological approach is necessary at all stages of control operations. Indiscriminate destruction of animals must be replaced by methods aiming to regulate the population levels of a pest species, with care taken not to involve nontarget species. Hitherto millions of useful animals have been unnecessarily destroyed just because it is simpler for man to kill en masse than to destroy selectively. This is bad economy.

Bats provide an example. Probably all New World bats are potentially vectors of rabies. Especially the sanguivorous vampire bats of Latin America are transmitters of this disease, which prevents local expansion of the cattle industry. Although livestock can be vaccinated against rabies, campaigns to eradicate the vampire bats by shooting, trapping, poisoning, and gassing have been carried out. Since vampire bats roost together with at least seventeen other species of bats, the control operations led to mass destruction of more useful species than of harmful species.

For many years Dr. Arthur M. Greenhall has been in charge of vampire bat control in Trinidad. In a paper published in 1968 he emphasized the general usefulness of bats to man. Insectivorous bats of both temperate and tropical regions consume enormous quantities of insects that without some agency to control their numbers would become overabundant. In the tropics, some bats serve as useful predators, feeding upon fish, lizards, birds, and mammals. Others, such as the long-tongued glossophagine bats that feed upon nectar, are essential pollinators for certain trees, while most of the fruit-eating phyllostomine bats propagate a number of economically important trees by the dissemination of seeds. In the temperate United States and Canada all bats are insectivorous and avoid contact with man and livestock. Therefore, indiscriminate killing of bats and destruction of their roosts are unnecessary and are harmful to human interests.

In 1967, a new federal policy for predator and pest control was initiated in the United States. In a "Statement of Philosophy and Policy for Animal Damage Control," the Department of Interior's Bureau of Sport, Fisheries, and Wildlife indicated their new approach: "Wildlife conservation must be practiced not only for the consumer, the sportsman—generally thought of as the hunter—but also for the ever-increasing proportion of people who simply enjoy seeing and hearing wild animals in their native habitat, or for that matter, simply enjoy the knowledge that these animals do exist."

Mortality, Overshooting, and Overfishing

The potential longevity of a wild animal species is greater than the average life-span actually attained. Death rates vary in different species. The rate of mortality is usually very high at low age levels, particularly prior to or soon after birth. In adult ages it is considerably lower. In contrast to the high death rate of newborn animals and juveniles, the mortality in adults of some species is surprisingly low. For example, in the gaur (*Bos gaurus*), a species sharing habitats with tigers, it is very low. A population study of this species in India showed an annual increase with a yearling percentage of only 10 percent (Schaller, 1967).

In birds that are not hunted by man, the annual mortality rate is usually between 12 and 30 percent, but in gallinaceous birds, which are frequently hunted by man, it is 60 to 80 percent (Farner, 1955). These figures come from the United States. The hunting pressure is also high on ducks and geese. The average life-span of 70 percent of the ducks in North America is less than one year. As has been stated, 71 percent of Swedish mallard (*Anas platyrhynchos*) deaths are caused by shooting.

However, the annual mortality is normally high also in "unhunted populations" of gallinaceous and anatid birds—due to various natural factors. Therefore, the secret of good management is to know what fraction of natural mortality can be replaced with harvesting by hunting. It is substantial but clearly variable!

For game animals all over the world the principal cause of death is undoubtedly predation by man. Therefore the mortality rate should be controlled by management measures setting kill limits that the game populations can withstand without declining in number.

Migrating geese and ducks need more than breeding grounds; they need moulting areas and resting places along their migration routes and winter quarters. Here the international aspect of the problem becomes obvious. Canada, the United States, and Mexico have shown how much can be done in collaboration by regulating shooting seasons and protecting migration routes and winter reserves. Europe has far more nations, and the obstacles are therefore many, but practical solutions can be arrived at in the Old World too, with a little good will.

By overkilling mammals and birds, man has added to the negative effects of other environmental destruction to cause depletion of animal populations all over the world. Most overkilling has struck game and predatory mammals and birds. In his drive on the predatory groups, man has thought that by attempting to wipe out "noxious" predators he would increase the stock of game. As a rule the effort has been wasted, for there has been little or no positive effect on the population of the so-called useful animals. The benefits of animal predation were not understood.

In lakes underfishing is as common as overfishing. Both should be avoided, because underfishing prevents a rational utilization of useful fish. Due to undernourishment resulting from competition, the fish do not attain normal sizes. Overfishing, on the other hand, leads to local depletions.

Because of rapid increase in human populations, increased standards of living, more leisure time, greater mobility, increased effectiveness of firearms and fishing techniques, the hunting and fishing pressures have increased tremendously all over the world. Simultaneously habitats and areas utilized by mammals, birds, and fish have become increasingly restricted and disturbed in many ways. Overshooting and overfishing in our time can be avoided only by regulations and strict law enforcement. Restoration of depleted animal populations must be undertaken through the creation of reserves and by management measures combined with a moratorium on shooting and fishing.

Coasts and Islands

Undisturbed coastal lands and islands are a condition of existence for many marine animals. Several of them are of great value to man—for example, marine turtles, many species of birds, and seals.

While marine turtles were dealt with in the chapter on the sea (pages 40–42), little was said there about the destruction of their habitats and the plundering of eggs and breeding females. The Caribbean was once the major reproduction area for turtles, but only a fraction of the former population remain in this region. The east coast of Africa, Madagascar, the Seychelles and other islands of the Indian Ocean formerly produced turtles in large numbers, but now they are rare in those places, and the turtle fisheries steadily decline. It is the same in southeast Asia, Australia, and the Pacific. In order to prevent the total destruction of a highly nutritive and agreeable food resource, it is necessary to establish a global restoration program of the marine turtles. Nesting beaches in all seas must be set aside as strictly controlled reserves where females can climb undisturbed to lay their eggs. These areas must be protected until the eggs have hatched and the baby turtles have reached the sea. The practice of catching egg-laying females on the beaches must be prohibited, and catches in shallow waters and the open sea restricted.

Carrying out such a program will be extremely difficult, but an effort must be made to save one of the world's most precious and irreplaceable food resources to counterbalance the present shortage of protein. Marine turtles occur in areas where the world famine is pronounced. It is also necessary to make use of the meat in a rational way. When exported as a luxury turtle soup, a large part of the meat's protein value is lost. Moreover, turtle soup is consumed in areas with an abundance of food. When the green turtle population has been restored and begins to yield large quantities of meat, turtle soup may again be marketed as a by-product of high economic value. In addition, eggs, skins, and oils are other useful products of all species of the highly valuable marine turtles. This global resource must be conserved, managed, and utilized wisely and efficiently for the welfare of mankind. Needed are international cooperation, scientific planning, and national protection of important coasts used as nesting sites by turtles.

The coastal habitats essential to marine turtles are just a fragment of sandy beaches above the high tide mark, where they deposit their eggs. Other animals need other kinds of shore habitat. Mammals of the seacoast include seals, sea lions, walruses, sea otters, polar bears, and sirenians. All of these species have high commercial value. Coastal birds play an important role in human recreation along seacoasts, and many of these species have high direct economic value for their guano, eggs, meat, skins, oils, and down.

Animal communities on isolated islands are of great scientific value for studies of evolution and speciation. Unfortunately, due to their high degree of specialization such isolated species are highly vulnerable to environmental disturbances. The endemic faunas and the habitats of the Galapagos Islands, Hawaii, New

Zealand, and other isolated islands have been altered drastically by introductions of exotic species. A great number of native species have thus become exterminated. Also, introduced diseases may wipe out animal species. This seems to explain the extinction of the endemic Hawaiian avifauna (Warner, 1968).

The scientific, cultural, and recreational values of islands are so great that even here a global conservation plan is necessary. A network of representative island reserves must be established before destruction has gone too far. Hitherto 111 species and subspecies of vertebrates have been exterminated during historic time on the islands of Oceania and on other isolated islands in the oceans. One hundred and twenty-eight species and subspecies of vertebrates of the same islands are at present threatened by extinction. These figures reflect the alarming rate at which islands are exposed to forces destructive of habitats and species.

Lakes, Rivers, Marshes, Bogs, and Other Wetlands

Among vertebrates, fish and amphibians that are considered to be economically unimportant are often regarded as being unworthy of protection. While the public conservation interest is chiefly focused on mammals and birds, the less observed fish and amphibians are neglected. This attitude is regrettable, because a great number of fish species are threatened by extermination through habitat destruction, alteration, and pollution. North America, Europe, Australia, and parts of Asia especially are in need of fish conservation programs. Native fish faunas must be given effective protection on a definitely greater scale than has hitherto been the case. Certain freshwater systems must be set aside as fish sanctuaries, where introduction of foreign species should be banned. More support and attention should be given to research on the biological and ecological role of fish presently regarded as being of no or little value.

Several species of fish have been exterminated through pollution, sewage effluent, or other oxygen-removing organic substances, and by introduction of exotic species. Chemical waste products and pesticides have in many cases contributed to the decline of species that are threatened by extinction. In several countries much attention and research is devoted to the artificial breeding of important native angling fish, which in their natural environment are then reduced by the impact of introduced species.

Overexploitation has also caused a serious decline in various species or populations. In Africa, for example, the spawning runs of some species have been reduced to fractions of their former abundance. Particularly mudsuckers (*Labeo*), noted for their migrations up rivers in flood to spawning sites in inundated areas, are vulnerable. During their migrations many thousands of mudsuckers are trapped at various localities along the rivers, where increasingly effective trapping methods are used. A tremendous toll is taken of breeding adults before spawning, and the populations thin out.

Artificial lakes for major water development projects alter the landscape and create ecological hazards that affect the lives of millions of organisms, sometimes even millions of people. Lake Kariba in Africa, the largest man-made lake in the world, provides a case history of disastrous detrimental effects on general productivity, including agriculture. Erosion, land shortage, famine, and human tragedies followed in the wake of the Kariba Dam construction and are now problems confounding the fallacious claims that had been made by technologists and economists. They planned the Kariba scheme, together with the politicians, without consideration for ecological factors. The lesson drawn from the mistakes made at Kariba shows the danger inherent in narrow, single-valued approaches—often commercially biased—that neglect the intrinsic complexity of the environment. In addition to the lake itself, enormous land areas around it and downstream were affected. For animals, including man, the scheme is a long-term catastrophe. Lake Kariba covers 1,718 square miles, but in a few years it will be second to Ghana's Lake Volta, which will flood 3,200 square miles.

The Kariba Dam project is just one example of shortsighted landscape development for unipurpose schemes without regard for detrimental by-effects. Other examples of damming of lakes and rivers with fatal environmental consequences are found all over the world. In all these cases it has been forgotten that natural hydrological cycles and animal life are renewable natural resources on which human populations often subsist. Even if the products of subsistence from a natural water system do not enter the gross national product of a country, because they are consumed locally, they play a vital role in the country's economy. Indeed, the subsistence output in undeveloped countries is in reality an important contribution to the national economy. Governments realize this fact in periods of famine when crops fail and the animal life is wiped out. Man cannot live on electricity alone.

The high density pothole country of the great plains of Canada represents about 10 percent of North America's total duck breeding area, but 70 to 80 percent of the continent's mallards and pintails are produced there. These two species happen to be the hunters' two most favoured duck species of North America. Moreover, the great plains wetlands are the only significant breeding region for a number of other species—for example, canvasback, redhead, and ruddy duck. Drainage of potholes for cultivation has reduced their wildfowl productivity a great deal. The importance of agriculture in these regions is indisputable. Therefore pothole drainage is increasing over large parts of the Canadian prairies and woodlands. However, the potholes are also of value to agriculture, a fact that is realized during drought years.

Stock raising is not antagonistic to wildfowl production. But the value of the prairie animal life itself must be considered and compared with other forms of production. The wetlands should be given multiple use.

In addition to the 100,000 or so prairie-dwelling humans hunting ducks and geese each fall for recreation and highly esteemed food, more than a million hunters in other parts of North America base their recreational activities on the ducks produced by the Canadian prairie wetlands. In addition, the prairie ducks, wherever they are, mean stimulating recreation throughout the year to millions of bird watchers. All these values and their economic benefits to business people not directly involved in hunting and bird watching must be taken into the balance. This is in fact done in North America, where waterfowl are publicly recognized as one of the United States' and Canada's most economically valuable and intensively managed living resources. Despite that recognition, North American populations of ducks and geese decline, due to agricultural drainage and encroachment. What has been said about Canada's vanishing wetlands is but an example. Less than a century ago the Canadian potholes and about ten million acres of the high plains marshes of Nebraska, Iowa, Minnesota, and the Dakotas produced 75 percent of all the wild ducks of North America. At that time there were about 200 million ducks, geese, and swans on the continent. At present, less than four million acres of wetland remain in the five states mentioned above. In 1968 the U. S. Fish and Wildlife Service estimated the breeding population of ducks (except scoters, eiders, mergansers and long-tailed ducks) to be about 32 million. Drastic drainage of wetlands has hit the most productive U. S. wildfowl country, for nearly two-thirds of the country's drained land is located north of the Missouri and Ohio rivers.

Ironically enough, governmental projects could jeopardize what federal agencies are trying to maintain or restore as productive wildfowl habitats. The proposed Rampart Dam on Alaska's Yukon River would have had disastrous effects on the wildfowl populations of the continent because it would have flooded 10,500 square miles of the greatest wildfowl nesting and wildlife areas of North America. Fortunately, the project was turned down when the findings of ecological investigations became known. The findings concerning wildfowl were summarized as follows: "The major losses of migratory waterfowl that would result from construction of Rampart Dam would be the complete extirpation of 1.5 million ducks and 12,500 geese that migrate annually from the Yukon Flats . . . Taking into account the efforts to date of all agencies and groups concerned with waterfowl preservation, the 1.5 million ducks produced on the Flats exceed the aggregate population of all federal and state refuges and marshlands restored by Ducks Unlimited and other non-governmental groups. In short, construction of Rampart Dam would negate thirty years of endeavour in waterfowl preservation in North America" (Leopold and Leonard, 1965).

Another major resource that would have been adversely affected by the Rampart Dam project was the salmon runs of the Yukon River. At present an average of 800,000 salmon are taken annually by local people to satisfy vital subsistence needs. This fishery has not yet been commercially exploited. Three species of Pacific salmon—chinook, coho, and chum—make the bulk of the runs, which are an enormous potential protein resource.

The geese and ducks produced in Alaska and the Great Plains region of North America emphasize that such habitats as marshes, bogs, and other wetlands must be regarded not only as ecosystems of immense local value but also as an essential part of a continental biome mosaic of benefit to a whole continent. In the Old World the wildfowl interactions are geographically even more complex, involving Europe, Asia, and Africa.

Even in the semiarid continent of Australia, with little and erratic rainfall, few large rivers and swamps, and minimal duck-goose habitats, wildfowl is likewise an important recreational resource. Apparently most of the Australian wildfowl species have adapted for survival through long drought periods. Like ducks and geese in temperate regions of the northern hemisphere that are extremely resistant to severe winters with long, hard frost periods and scarcity of food, their Australian counterparts seem to have a comparable capacity to withstand harsh conditions. In droughts small numbers of birds evidently manage to perpetuate the species in a few remaining swamps, and in wet years populations boom. But whereas wildfowl species of the world are traditional, with a regular breeding season and migration pattern, Australian waterfowl have to be opportunistic. In addition to being nomads, they breed at any time of the year as soon as environmental conditions permit. Their adaptation to extreme environmental conditions indicates their value as a resource. Despite the fact that Australia is the poorest of the continents in diversity of waterfowl, with only nineteen native species, the birds should not be ignored as a potential resource to be accorded conservation and management measures.

Tropical and subtropical lakes, swamps, and marshes produce and feed a great number of birds. They often occur in thousands upon thousands, and their number stands in direct proportion to the available food supply.

In addition to birds and fish, a number of mammals, reptiles, and amphibians contribute to the value and productivity of all kinds of wetlands—marshes, ponds, potholes, swamps, and flood meadows. These animal resources alone make many wetlands as economically important as farmlands and forests. Crocodiles play an important ecological role in tropical aquatic habitats and are of great economic value for their skins and meat. Yet most species have been seriously depleted and some are now threatened by extinction. In many rivers and lakes where the Nile crocodile, largest of the African species, has disappeared due to persecution, the fishery yield has decreased considerably.

Although the Nile crocodile preys upon animals ranging from crustaceans and mollusks to large mammals, during its various age stages its staple food in most localities is fish. Young crocodiles feed extensively upon invertebrates that prey upon the young of fish. Adult crocodiles feed chiefly upon fish without commercial value or on predatory species that prey heavily upon commercially desirable fish—among them *Tilapia,* which are the main producers of animal protein in African waters. Apparently the Nile crocodile maintains a balance in the fish populations by preying to a large extent on the carnivorous fish. It also preys on fish-eating birds and mammals. That destruction of the Nile crocodile does not benefit fishing interests has been found in several areas of Africa. After the reptile has been eliminated from a river or lake, the economically most important fish decrease in number and become increasingly mutilated in the nets due to attacks by carnivorous fish. The latter begin to dominate in the catches. This deterioration of the fisheries in areas where the Nile crocodile has vanished has led to conservation measures aimed at getting the crocodiles back. In several African countries crocodiles are now totally protected.

In Madagascar, larger crocodiles often preyed on unhealthy, weakened dogs. The elimination of crocodiles in Madagascar led to an increase of rabies, which is usually fatal to man.

An important investigation on the ecology of the Nile crocodile in Africa by Hugh B. Cott (1961) concluded that nothing justifies the classification of this crocodile as vermin. On the contrary, whether considered from the point of view of ecology or economics, it is a valuable and useful species. Probably other crocodiles of the world are likewise beneficial to fishing interests and of great economic value as producers of meat and skin. This is the case in Ceylon, where the reduction of crocodiles started a chain reaction that resulted in the disappearance of smaller insect-eating fish and an increase of malaria epidemics among humans.

Commercially, the most important food resource of the Amazon and Orinoco river systems was probably the South American river turtle (*Podocnemis expansa*). During the 1650's it was said that people living in the regions where this turtle occurred "never knew what hunger was." The turtle was so numerous that in some seasons it impeded the passage of canoes and smaller boats. This tremendous primary food resource has been utilized for more than three hundred years and was the basis of the economy along the rivers. The exploitation concentrated chiefly on the eggs, though the turtles themselves were hunted for meat and oil. The oil was used for lighting and as an edible fat. With no consideration for the future of the animal, eggs were collected in heaps twenty feet high. The reckless exploitation, combined with destruction of the shores where the turtle nested, caused such a decline in its numbers that by the middle of the nineteenth century it had become rare throughout its original range. Since then the river turtle has continued to decrease in numbers and distribution. Nesting is at present restricted to an area in the Trombetas River near Oriximiná in Brazil, where the species is still rather common, and to remote headwaters of the Amazon and the Orinoco, and in Bolivia's River Beni, Madre de Dios, and several other rivers, but even in these the species is threatened.

The Amazon manatee is another example of a valuable but ruined freshwater habitat food resource. Effective conservation and management measures could restore the population of this sea mammal to a level that would yield a considerable annual harvest.

Numerous valuable rodents frequent freshwater habitats—among them the coypu and capybara in South America, the musk-rat in North America, and the beaver in the northern hemisphere. Also, many protein-producing ungulates rely on aquatic habitats: the tapirs in South America and Asia, the hippopotamus in Africa, several species of antelope in Africa, the water buffalo in Asia, the moose in North America and Eurasia. Among aquatic carnivores otters, minks, and some other mustelids produce valuable furs.

On economic and social grounds, it is evident that perpetuation of the quality and quantity of animals produced by wetlands must be assured in the interest of human populations.

Deserts and Arid Lands

The Sahara is the world's largest desert. Most people regard it as a region that has been lifeless for eons of time. Actually, until comparatively recent times protein-producing game was plentiful in many areas of the Sahara. In the Late Pleistocene mighty and rapid climatic and vegetational changes reshaped large parts of the African continent, particularly the western Sahara. For a time extending into the last five thousand years the western Sahara, most of which today supports virtually no vegetation, must have been in sufficiently good condition for Mediterranean scrub and dry woodland to flourish all the way across to what is now its southern edge. On the other hand, at one stage the desert extended three hundred miles farther south than it does now. African vegetation limits have been very much on the move for a long time.

As recently as two thousand years ago elephants browsed the forest on the slopes of the Atlas Mountains. Hippopotamuses, giraffes, and rhinoceroses lived in the central parts of the Sahara around 2000 B.C. Crocodiles were then abundant in the lower Nile Valley of Egypt. As late as 1815 there were hippos in Egypt. Along with these animals a great number of grazing and browsing ungulate species, such as antelopes and wild asses, were widespread in northern Africa. At present many of them are gone, but fragmentary populations of scimitar-horned oryx, addax, dorcas gazelle, sand gazelle, mhorr gazelle, and Rio de Oro gazelle continue to have Sahara habitats. About one hundred years ago several of the species and some of the now exterminated ones occurred over almost the whole of the desert.

Competition with livestock and constant overhunting are depleting these useful wild ungulates, which in desert areas constitute man's best food resources. They have a phenomenal ability to survive in a harsh environment and to convert the extraordinarily poor vegetation into first-class meat, rich in protein.

The East and South African deserts also nourished large herds of wild ungulates until recent time, but almost everywhere man has destroyed or thinned out this resource. In the Nubian, Somali, and Ethiopian deserts the wild ass and several species of antelope and gazelle are threatened by extinction. At the same time the human population suffers increasingly from shortness of protein, and the livestock kills the poor vegetation on which it subsists. The Kalahari Desert in South Africa and Botswana continues to demonstrate the diversified fauna that deserts are capable of producing indefinitely because part of this area is protected as a national park.

The same kind of destruction is proceeding in Asian deserts from the Middle East to India and Mongolia, where many useful ungulate species have already been exterminated and others are doomed. They cannot be successfully replaced by livestock, which destroys the thin vegetation and has difficulty in surviving the hard climate.

Thus in virtually all the Old World deserts man is deliberately wiping out a formidable assemblage of large and medium-sized mammals that are useful food resources and remarkably well adapted to an extreme environment. This protein wealth produced by the poor desert vegetation represents the maximum productivity of the area. Man will never be able to produce more than a fraction of the former natural protein yield, and his dividend will diminish each year as his starving livestock kill their habitats.

Until recent times large mammals were also plentiful in the major deserts, semideserts, and arid lands of Australia and North and South America. The simple fact is that most natural arid lands of the world were initially productive in protein-rich game. Arid lands are usually thought of as inherently poor in economic values simply because people judge them by the present situation. They forget that most of the arid lands in the world today are man-made—once flourishing savannas or steppes that have been destroyed by human overexploitation in the course of hundreds, or even thousands, of years. Men, cattle, and goats are the primary makers of deserts. The true ancient deserts of the world are quite different from the false, man-made deserts. The steppes or savannas that are often adjacent to the latter show how deserts may develop if left undisturbed. Arid lands are not wastelands.

If the pressures from livestock, hunting, and other habitat disturbances were removed from semideserts and other types of arid lands, the faunal response in areas with a minimum of rainfall would be spectacular. It is a matter of management. Those in charge of freedom-from-hunger efforts should pay much more attention to the protein potential of deserts and arid lands when utilized by nomadic wild animals instead of plundering and destroying livestock.

A wise utilization of deserts also has a social aspect. While man has long been a desert inhabitant, only the bushmen of Kalahari and the aborigines of Australia presently use their desert environment without destroying it. Many nomadic tribes live in African and Asian deserts, where they chiefly subsist on sheep, goats, and camels. This way of life has recently been drastically changed in some areas, following the successful exploitation of oil deposits. In Arabia and in parts of the Sahara—notably in Libya and Algeria—the nomads, drawn into the oil industry by the increasing demand for labour, have settled down. The new social situation has both positive and negative consequences for the living natural resources of the desert. An example of the latter has occurred in Arabia, where gazelles and oryxes are hunted from cars. Unable to escape the massacre, these animals have been all but exterminated within a few years.

In North America the recreational values of desert regions are increasingly recognized. Large numbers of visitors come to the deserts and semideserts each year, attracted by plants, animals, and spectacular vistas. The displays of wildflowers released by the spring rains in the Mojave Desert of California and the Sonora Desert of Mexico are great tourist attractions. Moreover, these deserts produce high numbers of deer and other game animals, which increase their recreational values.

The general attitude toward deserts needs to be changed. The best way to use present deserts would be to utilize their natural yield, which in many African and Asian deserts is relatively high in the form of wild animal proteins. In exterminating the wild fauna by excessive hunting, and destroying wild vegetation by overgrazing of livestock, man is cutting his own throat.

Tundras

Although remote and climatically uncomfortable to human beings, climax tundra communities are astonishingly productive when not overexploited. Where wild reindeer and musk-oxen occur spontaneously, the tundra represents a wealth of high quality protein. Where the herds of semidomestic reindeer follow their masters or vice versa, the usual result is degradation of the habitat due to overgrazing, followed by deterioration of the domestic reindeer population.

Long-range management plans, adjusted to the carrying capacity of the vegetation, can repair this damage. But a question arises as to whether any human management, good or bad, is needed. It may well be asked if productivity from entirely wild reindeer populations is not higher than from domestic reindeer, without any investment other than for cropping, processing, storing, and marketing of the meat. In any event, it is certain that the return of wild reindeer herds, replacing the overlarge domestic herds, would help to restore ruined tundra habitats.

But what would then happen to the nomads who base their economy on domestic reindeer? In North America and Scandinavia, and at a later time in Asia, this problem can be expected to solve itself automatically. In Sweden, for example, the number of nomadic Lapps basing their economy on domestic reindeer is gradually lessening as an increasing number of young men and women settle in villages and towns and turn to a more sedentary and comfortable way of life—though the government tries to halt the process by various measures. This means that the herds of domestic reindeer will decrease in number and that probably no nomadic Lapps will remain in the not too distant future. Some areas in Lapland may support modest herds of "farmed" and fenced reindeer, but without the nomadic Lapps the wandering herds of domestic reindeer will inevitably disappear.

This development will create quite a new ecological situation, because the domestic reindeer have a tremendous impact on the montane vegetation. Alternatively, they may be replaced by wild reindeer, which means that there will be a return to the situation that prevailed for thousands of years during post-glacial time. In either case the change will be associated with the proliferation of the wolf and wolverine in the vast subarctic and arctic areas of Lapland. If the wild reindeer returns to Sweden and Finland, the ancient food chain of lichen-reindeer-wolf will be reestablished, providing the wolf has not been exterminated by that time. The wolverine and man will then be able to join the wolf as predators on the wild reindeer, which may develop into an important game animal, hunted for food, hides, and pleasure.

The question is, which will disappear first—the nomadic Lapps or the wolf and wolverine? It may be hoped that time will save the two carnivores, for it is not Utopian to imagine that the Europeans of the future will appreciate the plains, valleys, and mountains of Lapland, with its natural fauna of ungulates and predators, as one of the last of the larger wilderness areas in western Europe. Whatever happens in the northern tundras in the future, the conservation and management of the vegetation cover, particularly the nutritious and palatable lichens, will be a condition for productive life.

The caribou of North America and the reindeer of Eurasia, which were introduced into North America, are partly products of the tundra. Astronomical numbers of migrating caribou have been reported from Alaska and Canada in the past. The largest herd of recent time in Alaska occurred in 1921 in the Yukon-Tanana area with 568,000 animals (Murie, 1935). Reindeer were brought into Alaska in 1891–92 with the introduction of 187 animals. In 1932 they had increased to upwards of 625,000 individuals. Then the population overgrazed the lichens heavily and crashed, due to food shortage during the 1930–50 period, to about 25,000 animals. The reindeer competed seriously with the highly migratory caribou, which decreased, abandoned their old ranges, and did not come back after the collapse of the reindeer.

This example tells a lot about the danger of introducing exotic species and the vulnerability of slow-growing lichens, but it also illustrates the remarkable productivity of tundra regions. If conserved and managed properly, the tundras of the world can make an immense contribution of a high, sustained yield of ungulate meat, hides, and furs. This includes a potentially high yield from rodent populations—a resource that has seldom been utilized rationally by man outside the tropics.

In the present time the tundras of Alaska and other major

habitats of subarctic and arctic Alaska are facing the greatest threat they have ever had as a result of the discovery of what may be one of the world's largest oil deposits on the Alaskan Arctic coast. This region comprises some of the most valuable wildlife habitats and reserves of North America. The tundra and the taiga are the haunts of lynx, polar bear, grizzly bear, wolf, wolverine, moose, caribou, Dall sheep, swans, geese, ducks, and waders.

The exploitation of the oil and other mineral deposits has extended to arctic Canada. The Wildlife Society has asked the United States and Canadian governments to establish exploration guidelines for the protection of nonoil resource values. In addition the Society requested that consideration be given to the basic question of whether it is in the best interest of all the public—the rightful owners of the oil and gas reserves and other affected resources—to allow the industries involved to set the pace for the extraction of oil and gas, or whether it is more desirable from a long-range sociological, economic, and cultural standpoint for the governments representing the public to determine the most favourable rate of extraction.

In 1969 the U. S. Department of the Interior authorized the construction of a forty-eight-inch-diameter pipeline to carry oil across federal lands from Prudhoe Bay, on the arctic coast of the Beaufort Sea, to Valdez, a deepwater port on the Gulf of Alaska, eight hundred miles farther south. A set of stipulations will "insure that the wildlife and ecology of the Arctic, along with the culture and opportunities of Alaska's citizens, will be enhanced." One problem is how to insulate the pipe sufficiently to prevent the hot oil inside it from thawing out the permafrost, sinking into the soil, and so fracturing the pipe.

The tundra is a vulnerable habitat. Aside from the oil exploitation itself, serious damage will undoubtedly be caused by the great increase in human population and its activities.

Grasslands (Savannas, Steppes, Prairies, Pampas)

A number of sections of this chapter on the animals have already stressed the extraordinary animal productivity of natural grasslands in various parts of the world—especially in tropical Africa, where an assemblage of about thirty different species of wild herbivores can live in the same area together with large predators and scavengers. This is a formidable natural resource representing an extremely high degree of ecological adaptation, environmental balance, efficient land use, and sustained productivity.

At least half of Africa's land area is marginal or submarginal for crop production (Ledges, 1964). About 90 percent of Africa's lowland savannas are also marginal for domestic livestock production due to low or erratic rainfall. Most of these grasslands are extremely vulnerable to overgrazing and overtrampling, particularly during dry spells. In the tropics natural populations of wild ungulates seldom overgraze, while livestock almost always does so. Moreover, water needs, diseases, and other environmental obstacles constitute further limitations on livestock production.

The African human diet is in many parts of the continent deficient in animal protein. Domestic livestock cannot satisfy the protein needs of the continent's people for the reason that it is not generally adapted to the tropical environment and so destroys its productivity. The wild animals are not exposed to these environmental limitations. On the contrary, they are a product of the same environment. Therefore the utilization of wild animals as a food resource is the best and most rational use of African marginal lands, both for economic and ecological reasons. The results of experience all over tropical Africa clearly indicate that a new approach to land use planning, management, and utilization, based on wild animals, is necessary.

Of Africa's total area about 40 percent is occupied by savannas. About 37 percent is infested with tsetse flies, which transmit trypanosomiasis (nagana or sleeping sickness) to livestock, while wild animals are immune. This factor alone emphasizes the usefulness of wild animals in comparison with domestic stock. We shall come back to the tsetse problem a little later.

Much has been written about the enormous herds of wild grazing animals encountered by early European explorers of African savannas. The grasslands, intact and flourishing, produced game in quantities never found elsewhere by white men in historic time. A few centuries later, there are only traces of this fantastic living resource. In South Africa three hundred years ago the wild ungulate population totalled millions. In East Africa less than a century back many areas held incredible numbers of game. The situation was the same in West Africa, the Congo, Angola, Zambia, Rhodesia, and other areas of the interior.

Take the Kafue Flats in Zambia as one example. As late as 1921 sights like the following (Vaughan quoted by Robinette,

1963) were normal: "The fantastic quantity of wildlife defies description. Thousands, tens of thousands, and hundred of thousands of animals roamed over the Kafue Flats and the savannahs along its edges. . . . In travelling along the Flats from Mazabuka to Namwala and beyond you would never for one moment during that long trek have been out of sight of countless herds of game all around you, Buffalo, Eland, Roan, Wildebeeste, Zebra, Hartebeeste, Lechwe, Puku, Reedbuck, and the rest, while the savannah country bordering the flats was equally full of the browsers such as Sable, Kudu and Eland with the Buffalo using the dense musitu areas, lions were everywhere, the river full of hippo. . . ." In the early 1930's Pitman (1934) estimated there were 350,000 lechwe antelopes on the Flats. In 1960 the number was down to 25,000, but in 1971 it had increased to 94,000.

In terms of protein value, biomass, abundance, and variety of species, savannas and high veld are the most productive habitats in Africa. Undisturbed grasslands elsewhere in the world also have produced ungulate faunas of great value. This was the case a hundred years ago on savannas of India, steppes and high plateaus of Asia and Europe, prairies of North America, llanos, campos, and pampas of South America, and the grass plains of Australia. All these grasslands have been rapidly destroyed or greatly reduced within a few centuries. Many have deteriorated to desertlike appearance and produce only a fraction of their former yield. With domestic animals replacing wild animals, the result is a far-reaching habitat destruction that ultimately will lead to livestock suicide by starvation and an increased ecological crisis for man.

Extremely productive savannas in Africa have likewise been destroyed by man and his livestock. Only areas protected by tsetse flies or by national parks and nature reserves still carry dense populations of wild animals. In the arid steppes and semideserts of Asia and Europe the saiga antelope has convincingly shown how tremendously productive and economical relatively poor grasslands are when utilized by animals well adapted to the environment. Some of the few remaining natural savannas of tropical Asia still give evidence of their productivity. In Cambodia, in grasslands of the Mekong region, fifteen species of mammals can readily be seen, but this assemblage of precious protein-producing herbivores is now threatened by the Mekong irrigation scheme and the war.

In Australia the kangaroos, largest animals of the continent, are more efficient than sheep and cattle in converting vegetable food to animal protein, and their carcasses provide a considerably higher proportion of edible meat in the form of lean muscle. The kangaroo does not destroy its environment and it is particularly well adapted to live under marginal conditions. In contrast, grazing by sheep and cattle in Australia has led to a marked degradation of the natural pastures, particularly in the arid and semiarid regions. The old story about one kangaroo eating as much as eight sheep is unfortunately still widespread in Australia, where farmers believe kangaroos ruin their pastures. The following example shows how wrong this opinion is. In a one-acre hillside paddock seven large, grey kangaroos have been pastured for nine years. No soil erosion has occurred. The grass is as good as ever and even yields a crop of hay from time to time. An acre has been sufficient to feed seven large kangaroos throughout the year. The same number of sheep would turn such a paddock into a dust bowl within six months.

In a considerable portion of the world, grasslands of various type provide a potential protein resource based on wild animals that is often the only practical form of land use. A giant plan for restoring the world's formerly productive, game-yielding rangelands must be worked out and implemented before these areas have been irreversibly destroyed, and the wild animal resources lost forever. This task has to be a part of national land use policies and go hand in hand with a public education program to develop an understanding and appreciation of what conserving renewable natural resources means in practice. Conservation management and utilization of wildlife resources, such as grassland animal communities, can increase the protein productivity of a country and will yield other economic values such as recreation, education, and revenues from tourism. They can make the land soundly productive and as attractive as it can and should be.

Cultivated Lands

Animals may adapt to and settle in cultivated lands providing environments that meet their ecological needs. Many species of wild vertebrates are able to utilize cultivated habitats; others are disfavoured by or eliminated from them. In temperate regions such species as the European moose, red deer, fallow deer, sika deer, roe deer, and others are positively favoured ecologically by cultivated landscapes. That has been a factor, for example, underlying the tremendous increase in the European moose and roe deer populations in northern Europe. In Sweden alone, the meat value of the approximately 30,000 moose killed per year is more than $5 million. Most of these live in forests that man has developed or on lands with a mosaic pattern of woods and cultivated fields. In the tropics, secondary lowland rain forests attract a large number of forest ungulates. They find greater browsing possibilities in the thick undergrowth than in the thinner cover under the shadowing canopy of a primary forest. Therefore in many temperate, subtropical, and tropical areas wildlife production is an economically justified use of land that fits readily with other forms of land utilization. It should be a part of a general conservation and management policy.

Cultivated lands often change drastically. In habitats within an ecosystem disrupted by cultivation, the physical forces of the environment work chiefly under the influence of man, giving the animals little chance of making their own contributions to landscape changes. However, well-settled species—grazers and browsers, for example—definitely influence cultivated habitats, which then evolve in some degree independently of man. Even if man wishes to keep a stable cultivated environment with wild animals as a part of it, there is little possibility of maintaining the status quo. The alterations may be so slow that they are difficult to detect, but they are inescapable.

A large number of rodents, insectivores, and birds have become adjusted to cultivated habitats, since many of these species are dependent on grains, seeds, and insects. In the tropics such rodents are cropped and constitute an appreciated protein supply. A number of gallinaceous birds are produced by cultivated fields. Highly esteemed as game all over the world, they are economically important and should be regarded as a part of the land yield. Once again it is clear that a multiple use approach can provide more diversified crops and pleasures for cultivated lands than just monocultural harvests.

Forests

The value of forests as a timber resource is universally recognized. This has led to exploitation without regard for other forest functions, among them protection of steep slopes from erosion, conservation of water supplies, production of proteins in the form of wild animals. Forests are probably the best example for showing how the concept of multiple use of a habitat pays. If managed in an ecologically appropriate manner, forests can simultaneously preserve a water catchment area, yield timber, produce a regular protein supply, protect rare animals, and provide revenues from hunting license fees, perhaps also from tourism and recreation.

Although forests cannot compete with many grasslands as far as mammal biomass is concerned, they are in most cases very productive. Both temperate and tropical forests produce a number of large and medium-sized ungulates that could contribute immensely to the protein needs of the world. Unfortunately, most of the larger species of grazing and browsing mammals of tropical forests have been depleted, and many species have been nearly exterminated in the very areas where the human need for protein is greatest—Asia and South America. In Asia such large species as elephant, rhinoceros, tapir, banteng, gaur, kouprey, water buffalo, and many forest antelopes constitute a formidable potential of meat resource, but they have all diminished due to forest destructions and competition from less productive, habitat-destroying livestock. In India, particularly, domestic livestock are grazing inside the forests—a highly effective means of habitat destruction. The people of India fail to realize adequately that this destruction threatens human welfare. Where it is realized, government officials feel that it is impossible to control grazing by domestic animals in forests, or anywhere else, because India is a democracy where public property belongs to everyone. This idea is a direct route to catastrophe and collapse. The contrary concept is the proper one: precisely because it is public property, no one should be allowed to destroy a nation's most priceless possession, its fertile lands.

African forests produce a wealth of ungulate species ranging from elephants, buffaloes, bongo, okapi, greater kudu, smaller kudu, bush bocks, and two species of hogs to more than a dozen species of duikers and other smaller antelopes. These herbivores are only part of the protein resource of tropical forests. Primates, pangolins, rodents, birds, reptiles, amphibians, and many groups of invertebrates can be utilized for food. They are presently eaten in many areas of Africa, Asia, and South America.

Temperate forests too are important protein producers. Almost all the world's deer, including moose, are essentially forest dwellers. Bisons, beavers, an array of fur-producing carnivores and rodents, and many game birds are part of the forest life. In some cases careful logging in temperate forests may favour browsing mammals. In North America, removal of the forest canopy through logging has markedly improved the nutritive quality of plants growing on the forest floor. Likewise, some grazers and browsers in the tropics are favoured both for food and shelter by the undervegetation in secondary forests; which is denser than in primary forests with a closed canopy.

Montane forests throughout the tropics are disappearing rapidly. In addition to the serious economic consequences of this loss, many species and subspecies of animals with restricted distribution and specialized ecology are being placed in grave danger of extinction. Among the most critical areas of montane forest deterioration are parts of tropical America. A report (1965) from Colombia, typical as a reflection of what is happening in several places, expresses shock at the recent rapid deterioration of the environment. It states that the great Andean forests no longer exist except along the Pacific slope and on the Amazon basin side. The forests of the other slopes of the four Andean ranges are literally gone, as are those that until recently grew in the intervening valleys. Most of this destruction has taken place since World War II.

One tragedy of forest destruction is that many forest bird species do not go elsewhere or make comebacks. Most of them are unable to survive unless tracts of considerable size are left for them. When the forests go, the birds are gone forever. It can be prophesied with certainty that there will be considerable extinction of birds and other wildlife in Latin America within the next twenty years; indeed, some has surely taken place already, although specific information as to which creatures have gone is not yet available.

In the tropics the population increase, and in temperate regions the commercial interests, are the main reasons for the destruction of forests. The former problem is complex. The latter should not be difficult to solve. But all forest destruction is shortsighted, robbing posterity. Multiple use management based on ecological research can conserve the forest heritage and utilize its wildlife resources rationally on a sustained yield basis. It is absurd that the protein value of forest animals has been almost entirely neglected, or even regarded as a waste, in a world where about half of humanity is starving. Yet the animals are just a part of the utility of conserved forests. Time is running out. Decisive action for a global conservation program in the tropics and subtropics must be taken soon.

Migratory Animals

Many animals move regularly and seasonally back and forth between regions. Many birds shift between breeding areas and wintering grounds that may be located continents apart. These areas are essential for their survival, as are the feeding and resting places along their flyways. Marine mammals, reptiles, and fish also migrate widely in the oceans, utilizing quite different regions at different times. Many species of terrestrial mammals undertake migrations that may be occasional, seasonal, or nomadic, but in all cases are vital for their well-being. Anadromous and freshwater fish of many groups are migratory and cannot survive if their passage is cut off by habitat changes. Transforming a stream from fast-flowing rapids to deep, still water, or vice-versa, alters one ecoystem to a quite different one and so eliminates many migratory and sedentary species.

The environment of all categories of migratory animals is vital to their survival at all times, wherever they may be at whatever season of the year. Migration is, in fact, a selective process that in the course of thousands of years has enabled each species to work out its most efficient way of life. Of great survival value, it corresponds to the most economic utilization of the habitats the migratory species occupy seasonally.

Many of the migratory animal populations are of great economic value to man. Animals travelling across national boundaries cannot be managed or "administered" properly by a single government, unintegrated with management measures in other areas occupied by these animals. Thus countries producing or visited by migratory animals are responsible to all other countries that are temporarily frequented by these migrants. The attitude that the producing country has the moral right to do what it wants with its animal populations is wrong, both morally and economically. In the case of birds, for example, the winter areas and the feeding places during migration are of the same importance to the existence of the species as are the breeding sites. When one country drains a wetland that is important for nesting, feeding, or wintering wildfowl, it destroys property of other countries and people as well.

Therefore governments are responsible for cooperating with other nations on regulations to protect, manage, and utilize migratory species of wildlife and their habitats. Several such treaties already exist. The Migratory Bird Treaty act of 1916 between Great Britain (for Canada) and the United States is a model for other countries. A treaty between Mexico and the United States, ratified in 1937, provides even further protection. Ocean fisheries and sealing and whaling industries have also been regulated internationally, although these agreements do not always function properly. The whales, many of which are migratory, are an alarming example of shortsighted destruction of a global resource. Many inland fisheries are also internationally regulated. Lake Chad in Africa, with its tremendous protein resources, is a good example.

Species Threatened by Extermination

Once an animal species is exterminated, it can never be replaced. In our time man is the only exterminator on earth, and only he can save threatened species. The simplest, cheapest way to save them is to cooperate with nature.

However, not all animals are affected by destruction of their habitat. Many species have been able to increase their population and extend their range by adapting to man-made habitats and structures. For other species, alteration of the environment has gone beyond their ability to adapt. Almost all large wild mammals in the world today are losing out in the struggle for survival and are approaching the point of no return.

The situation is highly unfortunate for man because wild animals, wild vegetation, and living natural landscapes are an antidote for the sterility of technological civilization. Wildlife is the last reminder of man's wild past, teaching that we too are a dependent part of the biological environment. Man is ultimately as vulnerable to habitat destruction as are wild animals.

Within historic time—a split moment of the time vertebrates have existed on earth—man the plunderer has exterminated 360 species and subspecies of mammals, birds, reptiles, amphibians, and fish. Further, he has destroyed so much animal habitat that some 975 vertebrate species and subspecies are threatened with extinction. If they go too, the loss to human culture will be so colossal that future generations will look on twentieth century man as barbarians.

The whole civilized world became upset by the plan to flood the Nile's waters for the unnecessary Aswan High Dam project, which was going to inundate 3,000-year-old temples and monuments, nearly 50 Nubian archeological sites, and countless antiquities—not to speak of other ecological complications in the land of 122,000 Nubian villagers. A portion of these treasures were rescued by relocation before they became submerged, thanks to funds contributed by people everywhere. Thirty-six million dollars were subscribed to save these monuments. A fraction of this amount of money could save many animal species from extinction. Yet no organization or government has suggested a global subscription to save vanishing animal species, which are living monuments from ages before man existed. Each species is a master product of the environment, of an evolutionary chain representing an unbroken line of descent since the dawn of life. Its scientific value is priceless. Apparently man's concern for living treasures is less than for man-made ones. This is unconscionable conceit on man's part. It is dangerous to his future, leading as it does to apathy toward the continued lowering in the quality of man's living environment caused by technological pressures. The wounds inflicted on nature by extermination of species remain forever.

Animal species are far more precious than most people and governments seem to realize. The species that are threatened by extinction are not single, isolated kinds of organisms, to be regarded merely as scientifically interesting or esthetically attractive. They are an integral part of a habitat, of an ecosystem, of a complete living landscape where evolution is in full swing. Such ecosystems, intact with a full set of animal species in action, may serve as sample areas providing invaluable information as guidelines for developing other areas without destroying environmental productivity and health. The genetic importance of preserving animal species in a natural system must be stressed. Only in free-living populations of a species is a free exchange of genes realized. No one can know in advance what environmental, chemical, or medicinal value a wild species may have for humanity in the future, just to mention a few economic justifications for their preservation.

Other good reasons of a less material character exist for the preservation of threatened species. An important part of man's increasing interest in conservation is motivated by his appreciation of aesthetic values and by ethical principles. Because man alone is guilty of vandalizing the earth and its animal life, many people feel strongly that the present generations are responsible as custodians of a heritage that must not be further destroyed. But these people are still a minority.

It can be anticipated that during the decades to come increasing numbers of humanity, modern technology, and proliferating urban agglomerations will further encroach on land and water that are now fully or partly utilized by wild animals. Simultaneously man's protein hunger and profit greediness will increase the rate of animal extinction. The human pressure on the rare species, either by direct killing or environmental overexploitation leading to habitat destruction, will be tremendous. Before it is too late a global rescue operation to save the world's rare animal life must be organized. Abu Simbel's ancient temples at Aswan could be reborn. Wild animal species cannot.

The main causes of the decline and ultimate extinction of animal populations in the last centuries are destruction of habitats and excessive cropping by man. Human overexploitation, too often due to greediness, is the primary reason for the decline or extinction of all wild animal populations of any economic importance in historic time.

In the geologic time-scale, animal species will become extinct regardless of man's effort to preserve them. But at least 99 percent of the animals now threatened by extinction are in this fatal situation due to man. Hunters have given the *coup de grâce* to many species by reckless, continuous overkilling. This form of extermination rarely occurs for subsistence purposes. Rather it is a phase of the commercialization of animals as food, oil, hides, furs, skins, feathers, horn, ivory, and other kinds of products. The demand in Asia for aphrodisiacs in the form of pulverized rhino horns is leading to extermination of the world's five remaining species of rhinoceroses. The demands of medical research have thinned out ape and monkey populations to such low levels that several species are threatened with extinction. Predatory mammals, birds, and reptiles have been killed in enormous numbers because they are wrongly believed to be antagonistic to human interests.

Enforcement of legislation designed to protect threatened species can scarcely be an effective tool unless it is carried out hand in hand with education and informational programs that teach the ecological role and economic value of wild animals as a renewable resource.

Trade in Wild Animals

Trade and transport in wild animals for various purposes have increased in recent decades to the point where they constitute a threat to many species and regional populations. The pet trade is uncritically and irresponsibly draining wild animal populations. The demand for apes and monkeys from centers of medical and pharmaceutical research is so heavy that it now constitutes a menace to the living supply. The marketing of fashion furs from spotted wild cats like leopard, snow leopard, cheetah, and ocelot, and from aquatic mammals such as seals and otters, as well as skins from various species of crocodiles, lizards, and snakes, and such other animal products as ivory, rhino horns, and so forth, has led to local exterminations and serious depleting of animal populations. Tourist hunting-safaris, irresponsibly organized, are also a threat to several species—among them the polar bear, the walrus, the tiger, and the jaguar. The whaling industry is a monstrous instance of reckless commercial plundering of a global resource.

Much of this human consumption of wild animals is totally unnecessary and meaningless. Exterminating animals just for the fashion business, or to satisfy animal dealers involved in pet trade or people who want to kill rare animals for fun, is inexcusable. That so little public indignation is generated against these moral crimes must be due to ignorance of what crass commercial trade is doing to wild animals. In 1967 alone more than 28 million live wild animals and more than 22 million pounds of wild animal skins were imported to the United States by American pet dealers, laboratories, zoological gardens, furriers, and leather goods concerns (W. G. Conway, 1968). In 1968 the figure of live animals imported to the United States had increased to 67 million and in 1970 to 86 million. A tremendous proportion of animals transported or kept by animal dealers and eventually later sold to private individuals suffer and die because they are inadequately cared for by unqualified people without elementary biological knowledge of animal needs.

Complete records of imported live wild animals into the United States are available for 1967–1970. They are as follows:

	1967	1968	1969	1970
Mammals (all)	74,304	140,858	116,341	101,302
(Primates	62,526	124,440	99,668	85,151)
Wild birds (other than parrots & canaries)	203,189	492,280	571,663	687,901
Fish	27,759,332	64,254,190	73,694,996	83,867,029
Amphibians	137,697	170,621	339,489	572,640
Reptiles	405,134	1,950,091	1,393,970	2,101,751
Total	28,579,656	67,008,040	76,116,459	87,338,623

These figures are alarming and show a tremendous increase in only one year (1967–68) for every group of animals, while in 1969 the import of mammals and reptiles decreased. For mammals this trend continued in 1970, but for reptiles the figure increased. Birds and fish are imported in rapidly increasing numbers. The figures for 1970 are three times higher than for 1967.

The trade in many primates is based on babies, or very young individuals, that as a rule can only be captured after the killing of their mothers. A shockingly high percentage of captured animals of most species die during or following the capture operations. In one instance in 1968, for example, a group of mountain tapirs were brought into Quito for export after sixty others had been killed in the course of capture. Moreover, great numbers of captured animals perish during transport. It is not unusual for 75 percent of small birds in a shipment to be dead on arrival.

Additional large losses occur at the next stage, with the animals in the hands of irresponsible dealers who base their business on the hope of selling their charges quickly before they die. Finally, most pet owners have very little idea of how to take care of animals, many of which are delicate and soon die. In the United States the trade in horned lizards (*Phrynosoma platyrhinos*) has been going on for decades, and they are even for sale in five-and-ten-cent stores (Conway, 1968). These lizards almost invariably starve to death after a few weeks in captivity. This well-known fact among animal dealers does not stop them from continuing their pet trade in horned lizards. There are a great number of other delicate animals that likewise die in captivity after a few days or weeks and therefore are completely unsuitable as pets, yet the pet dealers continue offering them for sale.

The United States, the Netherlands, and Great Britain seem to be the main countries importing wild animals. The trade from South America, chiefly to the United States, reaches astronomic figures. One of several recent shipments of small birds from Colombia to New York contained 28,000 birds. According to Professor Carlos Lehmann of Colombia, ten times that number die before shipment. He also estimates that due to poor shipping methods fifty birds die for every one that reaches its destination. More than 39,500 live animals were exported from Peru in 1964. Of these, 26,200 were squirrel monkeys and 6,300 were other small monkeys. The 28 million vertebrates imported alive into the United States in 1967 included 62,500 wild primates, imported mostly for research purposes. In some previous years wild primate imports exceeded 200,000. In 1968 between 160,000 and 200,000 nonhuman primates were involved in world trade, according to figures presented at the Third International Congress on Primatology in Zurich in 1970. These apes and monkeys are used for medical research, by the pet trade, and by zoological gardens.

The species most threatened at present by the pet trade is the chimpanzee, of which about 6,000 (allowing for losses) are being taken every year from the wild.

Practically all crocodilians (alligators, caimans, crocodiles, and gavials) of four continents are continuously persecuted and depleted for commercial purposes more rapidly than they can reproduce. This exploitation of crocodiles during the last few decades has led to local extermination of many species in most areas of their ranges. In India, for example, the scale of destruction can be gauged from the fact that in 1948 and 1949, 16,000 and 30,000 skins respectively were exported through the port of Calcutta alone, and the estimated production of crocodile skins in India was 40,000 to 50,000 per annum. In 1967 the export of crocodile skins was banned in India, but none of the three species occurring there is protected and the depletion continues. In Thailand the situation is the same, and there the Siamese crocodile, a fourth species, is now virtually extinct in the wild. Some specimens are kept in captivity. Of all crocodiles only the American alligator, after having been poached and slaughtered for years, is now increasing in numbers due to careful management.

From Morocco some half million tortoises are exported every year to Great Britain alone, and similar numbers to other countries. This is an extremely heavy drain on the natural supply of several species.

The expansion of the pet trade is alarming. It should be regulated by national legislation and an international convention. The International Union for Conservation of Nature and Natural Resources (IUCN) has been working on an international convention concerning trade and transport of wild animals. The draft is nearing completion as this book goes to press.

However shockingly large the figures on export and import of live animals for the pet trade may be, they are lower than those for commerce in dead animals in the form of furs and skins. During the period 1951–60 skins and hides of at least 7,669,758 animals were exported from Argentina, in addition to more than 28,000 live animals. Among the dead animals were 5 million iguanas and more than 100,000 rheas. Bolivia alone exported five tons of vicuna wool to the United States in one year, which represents the deaths of between 25,000 and 30,000 vicunas. In 1961 skins from 26,000 colobus monkeys were shipped out of a single Ethiopian port. In the same country the leopard has been locally exterminated because its fur commands a high price in countries where fashion creates a demand for coats of spotted cats. The ocelot is another example of this vanity. From Peru alone 11,244 ocelot skins—out of a total 247,956 animal skins and hides—were exported in 1964. The world catch of spotted cats in 1967–68 is estimated at half a million animals, of which about 350,000 were imported to the United States. Of these about 200,000 were ocelots and 20,000 were leopards. The import figures for tigers was 200. Even snow leopards were imported to the United States. Total imports of the most important "fur species" of spotted cats to the United States alone in 1968, 1969, and 1970 were as follows:

	1968	1969	1970
Leopard	9,556	7,934	996
Jaguar	13,516	9,831	7,758
Ocelot	128,966	133,069	87,645
Cheetah	1,283	1,152	
Total	153,321	151,986	96,399

In 1967 the United States imported nearly 22 million pounds of furs, skins, and hides, not including the trade in finished fur, hide, and leather products. These imported skins and hides represent a wide range of species. 970,809 deer and antelope hides were imported from South America and Africa; 115,458 ocelot, chiefly from Brazil, Bolivia, and Colombia, and 35,748 otters from eleven Latin American countries (Conway, 1968).

The International Fur Trade Federation has agreed to recommend that its members totally ban trade in skins of tiger, snow leopard, and clouded leopard. A three-year ban has been recommended on leopard and cheetah skins. This step by the furriers does not constitute an economic sacrifice, because less than one percent of the fur industry's $300 million in annual sales comes from spotted cats. The Federation is joining IUCN in a survey of the status of ocelot, jaguar, and margay cat in South America. Another survey will check the position of cheetah and leopard and other depleted species in Africa. The Indian, Pakistan, and Nepal governments have banned all trade and export of tiger, leopard, and other endangered cat skins. The tiger has become protected in all three countries.

In principle there is nothing wrong with utilizing the surplus of animal populations for human needs if the animals can withstand the cropping pressure and the killing technique is acceptable. However, most people buying furs do not need them. It is the hunters that need the money they can make from killing and skinning animals. The fur trade should not be tolerated when it is based on species or populations that are near extinction or are endangered by the trade itself. Nor should it be permitted when it is deliberately plundering useful resources in one area after another, even if the species concerned is far from being exterminated in other parts of its range. Most furs and skins are exported from undeveloped countries, where animals producing these products should be regarded as useful resources that must be conserved, managed, and wisely utilized so that they give a sustained yield. Conservation measures must take the place of blind slaughtering before it is too late to restore a precious capital resource.

At the end of 1969, the United States took an important step towards the protection of endangered animals in the world by passing the new Endangered Species Conservation Act. This bill governs the importation of wild animals and their products. The regulations are designed to put a stop to the importation of endangered animals into the United States, and to prevent the entry of wildlife species and products, which have been poached and illegally exported from the country of origin. This law not only protects native species threatened with extinction but all the threatened species of the world, whether they can be obtained legally in their country of origin or not.

The Act requires the Secretary of the Interior to keep an up-to-date, scientifically compiled list of endangered species, to authorize research and propagation programs, and to acquire habitats of up to a value of one million dollars a year. Appended to the law is a list of foreign endangered species threatened with extinction, the importation of which will be permitted only for scientific purposes or for propagation in captivity to perpetuate the species. This list is based on material available to the Survival Service Commission of the IUCN. The Secretary of the Interior will keep this list flexible through consultations with the IUCN and the countries of origin. Enforcement of the ban has been greatly simplified by allowing the import of wild animal species and their products into the United States only at certain designated ports.

The Secretary of the Interior is also directed to make positive efforts to get other countries to agree to conserve their endangered species and is empowered to give them technical assistance.

Legislation to prevent trading in animal species within the United States' own territory has been considerably improved and now covers reptiles, amphibians, mollusks, and crustaceans as well as the mammals and birds already listed. Previously some of these species were protected in one state but not in another. However, now it is illegal to trade, throughout the entire United States, in any animal species formally protected in only one particular state. This will constitute an effective barrier to poaching and smuggling.

By this new legislation the United States has again shown its global leadership in conservation and has set an example to be followed by other countries. It is to be hoped that the international convention of trading in wild animals, suggested and prepared by the IUCN, will soon be finalized as a result of the United States example.

There is little justification for a government to ignore the pleas of other countries and international organizations and to permit its citizens to be knowing receivers of stolen property. Too many governments have not yet taken action to prohibit the import of certain threatened species, despite requests to do so from the countries of origin. Reaction to the request of the Peruvian Government to ban the importation of vicuna wool has been slow, and indifference, greed, or negligence is evident. So far only the United States and Great Britain have followed the wishes of Peru to ban importation of vicuna skins and wool.

At the IUCN's General Assembly in New Delhi in 1969 a resolution, unanimously adopted, urged "all countries to uphold the principle that any wild animal or product taken in violation of one country's laws should be declared contraband in all other countries."

In this connection, the fact must again be stressed that responsible conservation organizations, like UNESCO and the IUCN, as well as the United States Government, are encouraging the rational use and management of wildlife resources, rather than preservation as such and non-use of these resources. Strict protection is necessary only when species are endangered by extinction, or inside ecologically autonomous national parks and reserves. Economic incentive can be one of the strongest motives to protect animal species and their habitats, in order to insure future abundance and rational utilization outside national parks and equivalent reserves. Scientific, educational, and aesthetic motives are also important but probably inadequate, in present times, as convincing arguments for proper conservation measures.

Pesticides

The chemical war against nature has been carried forward on an increasing scale for about half a century. Prior to World War II most pesticides were nonpersistent organics existing in natural environments, such as pyrethrins (pyrethrum), which was derived from plants. Since the war, the use of pesticides, or biocides (lifekillers), has attained tremendous proportions. A great part of this war against living organisms is directed toward animals, chiefly insects. Pesticides have undoubtedly contributed immensely to crop protection and to human health and welfare, but, on the other hand, they have also damaged useful plant and animal species, habitats, and ecosystems, and done harm to human health and welfare. Contamination of the environment with toxic chemicals like chlorinated hydrocarbons and polychlorinated biphenyls has gone so far that in many countries, among them the United States, Great Britain, and Sweden, DDT and other pesticides are found practically everywhere—in the air, in fresh water and the sea, in the soil, in plants and animals. Similar conditions hold in many other countries, some of which do not seem even to be aware of them. We have already added so many poisons to the world's ecosystems that it may be too late to restore and clean what has been damaged, among other things the food-producing capacity of the earth. In fact, of the four basic ingredients of life on earth—sunlight, air, water, and soil—all except sunlight are polluted with one or more organo-chlorine biocides.

DDT is the cheapest of all modern pesticides, and so it is the most frequently used. In twenty-five years DDT has contaminated the atmosphere, soil, streams, lakes, and the sea, and has infiltrated the tissues of most living organisms. A chlorinated hydrocarbon, DDT persists in the environment for at least ten years, is nearly insoluble in water, and cannot be dissipated in the oceans. However, it is extremely soluble in fatty tissue and builds up quickly in the fat of vertebrates. It readily attaches to minute particles that are swept aloft by winds or rainfall or transported by flowing water. It moves in food chains from one organism to another. It passes from mother to offspring.

The use of DDT and numerous other toxic chemicals is in many ways biologically wrong because in a general sense it signifies a dead end. Without knowing the capacity of man to endure the effects of DDT, we destroy environmental niches that may be vital for numerous other species. We distribute virulent poisonous compounds in such quantities that their side effects are no longer controllable. The long-term environmental damage caused by the use of many pesticides must be far greater than the limited short-term benefits. Therefore DDT and other poisonous compounds must be banned without delay or at least be restricted and regulated in a way that they cannot escape into the environment. This means that the use of persistent biocides in outdoor areas must be outlawed.

Pesticides often kill blindly, destroying not only the noxious insects against which they are used but also thousands of non-target insects of other species sharing the same environment. Among them are predators that function as a biological control of the populations man attempts to eradicate through pesticides. These predators reproduce much slower than the "pest" species, which soon compensate for their population losses, partly as a result of the elimination of the species that formerly controlled them. In fact, the most serious outbreaks of pests often occur immediately following, and as a direct consequence of, the application of contact-acting insecticides. Many convincing examples of this phenomenon have occurred in the tropics. Notably in agricultural ecosystems of Malaysia, it has been found that the more frequently pesticides are used, the more serious are subsequent pest outbreaks (G. R. Conway, 1969).

Moreover, the pest species develop resistance to the chemical compounds devised to kill them. During the last twenty years at least two hundred species of pests—mainly insects, ticks, and mites—have become resistant to pesticides. There are also indications of resistance among rodents and plants. The gradually developed immunity of pest species to pesticides leads to an increase in the dosage given at each treatment and to the marketing of increasingly powerful toxic chemical compounds. Stronger and stronger poisons have killed more and more useful insects and other animals in a vicious spiral of destruction with the risk at the end that the pests may be still there and that ultimately man himself will be hit.

The general result of the continuous dispersal of lethal pesticides in nature is a lowering of the environmental quality, with adverse effects on man. Often these chemicals are highly persistent. They accumulate in living organisms, in the soil, and in the water.

They are active long after their usefulness as pesticides, aimed at destroying certain pest species, has passed, remaining in the environment as a constant threat to living organisms.

Dimond and Sherburne (1969) substantiated the persistence of DDT in wild mammal populations in the United States. They found residues of DDT in moles, mice, and shrews eight and nine years after single applications of DDT in forests at the rate of one pound per acre to control the spruce budworm. The mice and moles contained small but significant quantities of pesticides up to eight years after spraying, while the more predatory shrews carried ten to forty times as much. Some of these shrews were living in forest areas nine years after a single spraying and must represent several generations.

The degree of danger to which these small mammal populations are exposed by the residue levels they carry is not fully known, but it is certain that the residues constitute a real hazard to predators that prey heavily on them. This is analogous to man's dilemma.

Despite their accumulating effect, pesticides are used indiscriminately in many parts of the world at a constantly increasing rate as a conventional killing tool, with little or no concern that much more is killed than man wants or needs to kill with it. In 1966 in the United States, according to the Department of the Interior, the use of toxic chemicals had increased to the point where one out of every ten acres was being sprayed with an average of four pounds per acre per year. In 1967 more than 80,000 pesticides were registered in the United States. In 1968, 1.2 billion pounds of synthetic organic pesticides were produced there, of which about 20 percent was exported. In 1969, Senator Gaylord Nelson of Illinois told a U. S. Senate committee that high concentrations of DDT, substantially above the five parts per million set by the U. S. Food and Drug Administration, have been found in fish in almost every region of the country.

Slightly more than 50 percent of all of these pesticides were used in farming. As early as 1968 it was estimated that about one billion pounds of DDT were circulating in the biosphere. Today hardly any area or living organism in the world is untouched by man-made chemicals. Even marine environments far away from developed areas and deep in the sea are affected by DDT. There is a high concentration of DDT in polar bears in the Canadian Arctic and in penguins off Antarctica, not to speak of seals and salmon in the heavily polluted Baltic Sea.

Inasmuch as pesticides are in the long run a threat to man as well as to plants and animals, and may even drastically reduce the oxygen production of our planet, the attitude of a Food and Agriculture Organization (FAO) working party on pesticides comes as a shock. It stated in 1965 that the resistance of animals to pesticides "is becoming a real threat to the continuation of the use of these chemicals as tools in food production and protection"; however, after surveying the availability of other toxic chemicals, it noted that prospects for a continuing supply of alternative pesticides were good. This is a typical technological approach of our time. The specialist does not see beyond a very narrow niche. His aim is to destroy a certain species. Simultaneous destruction of the world around that species is incidental.

In fairness to some of the suppliers of agricultural chemicals, it must be said that they admit that wanton misuse of pesticides would inevitably carry their effort far beyond what was intended by proper application. But what is "proper application?" One can never strike blindly with a weapon without causing harmful side effects. It is strange that the increasing resistance of pest animals to toxic chemicals, along with other disadvantages that have become apparent, has not yet created doubt in the minds of manufacturers and distributors of pesticides, and administrators responsible for application programs.

Governments and manufacturers have unrealistically and irresponsibly minimized the dangers involved in the use of pesticides. For more than two decades conservationists have been talking about the problem of pesticides and environmental pollution. Rachel Carson and others were accused of greatly exaggerating the seriousness of the detrimental effects of toxic chemicals, which Governments and manufacturers of poisons justified as the necessary side effects of technological activities that had to be accepted in the name of human progress. In 1950 the Tenth World Conference of the International Council for Bird Preservation, held in Uppsala, Sweden, adopted a resolution that urgently warned governments about the use of toxic chemicals. Not one government took action at that time or showed any concern over the warning. In 1965 a voluntary agreement was reached to ban particular uses of such toxic chemicals as aldrin, dieldrin, and heptachlor in Great Britain, out of recognition that these compounds were too damaging to the environment, and particularly harmful to vertebrate animals. This voluntary ban was a fine achievement, but when the British market was closed, the same manufacturers tried to introduce their harmful toxic chemicals in other countries. Some years later, when comparable Swedish products were banned in their home country, the Swedish manufacturers did the same. Now toxic chemicals banned in several countries of origin are dumped in developing countries in Africa, Asia, and Latin America. It is criminal of governments to allow export of products which for safety reasons are not permitted for use in their own country. Many of these toxic chemicals are now used in the tropics, despite the fact that application of non-selective, persistent pesticides under tropical conditions in some cases aggravates the attack of parasites.

In Tanzania alone thirty-six people died of insecticides within the period July, 1969–July, 1970. Postmortems gave positive evidence the victim had eaten food mixed with insecticides. The government chemist believes that if all the deaths caused by insecticides had been reported, the number would have been even higher.

In the United States each year between one hundred and two hundred human deaths are known to be caused by pesticides.

In Honduras thirty-seven soldiers died in 1971 because of food poisoning after their barracks had been sprayed with an insecticide.

Laboratory studies of mammals like primates, dogs, and rodents have shown that some pesticides increased the frequency of birth defects, genetic mutations, and cancer.

It is often claimed that pesticides are a necessary adjunct to the production and storage of food vital to the people of the world and for control of diseases afflicting man. Pesticides have rendered important service by increasing crop yields and by locally eradicating disease vectors, but the problem must be handled in the larger perspective of its relation to the future of living environments and ecosystems. Even conceding that the well-being of man on this earth is paramount, it does not follow that man is well served by using techniques that keep human populations going today and will smash them tomorrow. Man is threatened with the dire likelihood that the toxic chemicals now helping to feed him will eventually produce a cumulative poisoning of his environment that will create his graveyard.

Manufacturers of toxic chemicals claim that their products "must logically be seen as a necessary evil in the struggle to feed mankind." They also say it is regrettable that the uses of pesticides "occasionally have side effects involving, perhaps, a temporary reduction of certain species of wild life." This is a deceptively mild way of stating the case. Several vertebrate species have, in fact, been exterminated in areas where pesticides have been used. Soil bacteria and fungi so vital to life are often destroyed by persistent pesticides in the soil. The side effects are of such magnitude that man himself is among the threatened species. The time is at long last at hand when alerted officials are being forced by public pressure to outlaw the sale and use of DDT and other persistent pesticides. In fact, this has already happened in some countries (Australia, New Zealand, Canada, Cyprus, Hungary, Norway, and West Germany) and some states in the United States. The U. S. Secretary of Agriculture announced in June, 1970, that he did not intend to suspend sales of DDT in the United States, while the Secretary of Interior banned all uses of DDT, aldrin, dieldrin, endrin, DDD, and mercury compounds, and other toxic chemicals in all public lands managed by his department.

Denmark has stopped agricultural use of DDT. Sweden has banned DDT but allows its use restrictively in forestry. In Switzerland DDT will be banned after 1972. Japan has decided to ban DDT and limit the use of BHC. These steps signify local progress, but nevertheless the annual production of DDT amounts in 1971 to between 200,000 and 250,000 tons.

As mentioned in the chapter on freshwater environments, aquatic organisms rapidly absorb chlorinated hydrocarbon pesticides like DDT. Polychlorinated biphenyls (PCB) are apparently also found in ecosystems in large amounts. The recent findings that pelagic seabirds dying on the coasts of Great Britain had these toxic chemicals in their livers in concentrations of several hundred parts per million (Bourne and Mead, 1969) are evidence of this. Also, there are high values of PCB in fish of the Baltic Sea. We still know very little about these persistent chemicals in the environment. Polychlorinated biphenyls are complicated because they do not represent a single compound. Their effects are long-ranged in mammals and birds and particularly cause liver alterations, but the toxicology is rather poorly known in comparison to the present knowledge of chlorinated hydrocarbons.

Residues of PCB have also been found in freshwater and marine fish, so it is logical to assume that fish-eating birds are contaminated. In the aquatic food chain these birds represent the same final link as man and other mammals. The recent identification of PCB in waters and sediments has raised a question concerning the movement of these toxic chemicals through the food chain.

The diversity of these residues and the inadequacy of analytical methods for their determination have prevented a thorough evaluation of this problem. Many industries used PCB, and their dispersal throughout the world likely results from aerial contamination and transportation. Since PCB components are stable, insoluble in water but highly soluble in lipids, it is inevitable that they become concentrated in biological systems. The chemical structures of PCB residues are similar to those of several organochlorine biocides, and it is expected that their movement through biological systems would be similar (Chesters and Konrad, 1971).

Mercury compounds, which have been used extensively in Sweden, the United States, and probably elsewhere but still undiscovered, are likewise quickly absorbed and stored in the fat of fish and further on in other links of the food chain, man included. In Great Britain, where conservationists—particularly ornithologists—have made revealing studies of the effects of pesticides on wildlife, a number of significant facts have emerged. Birds that feed on fish or other freshwater food have shown the highest average residues of organo-chlorine compounds among those analyzed. The second highest concentrations were found in birds of prey and owls. Since these birds prey selectively and, moreover, are at the end of the food chains, they rapidly accumulate dangerously high quantities of poison. Of 236 bodies from 84 species examined, only 4 percent were completely free of organo-chlorine residues, and none of 145 eggs of 34 species was free of contamination. Analyses of insects and other invertebrates showed that all were contaminated. One dragonfly had residues of four different organo-chlorine compounds.

Nontarget organisms are often hit, but such accidents are usually recorded only when they come in direct conflict with man's interest. In fact, most of the volume of pesticides does not reach the target at all. An example is what happened in a coniferous forest in Swedish Lapland. On July 12, 1969, herbicides were sprayed by airplane. In November, 1969, a Lapp family arrived there with about 600 reindeer. The Lapps had been directed to the area by the governmental authorities. In April, 1970, the reindeer started to eat leaves. Within a few weeks nearly 250 reindeer died and 40 animals had abortions. The foeti looked unusually premature and had probably been dead in the uterus for some time.

Spraying with DDT for Dutch elm disease in New Hampshire caused the elimination of 70 percent of the American robin

(*Turdus migratorius*) population. Similar instances are reported for Connecticut, Michigan, and Wisconsin. Many even worse examples of negative effects on nontarget organisms are known. In Florida during the 1950's a spray operation using dieldrin against sand flies killed 1,117,000 fish of at least thirty species totalling about twenty-five tons. Crustaceans were virtually exterminated. In southern areas of the United States, where dieldrin and heptachlor were spread aerially over two and a half million acres in 1957 to eradicate fire ants at a cost of $2.4 million, the operation took a heavy toll of vertebrate populations. Some did not seem to have recovered seven years later. Ironically, it was subsequently found that the fire ant is not really a pest species in an economic sense and can be controlled by much less costly means, with few side effects.

Aerial spraying of agricultural crops and forests increases every year on all continents and is one of the most serious forms of environmental pollution by toxic chemicals, particularly in aquatic habitats. Conditions for aerial spraying are seldom perfect. It has been estimated that even in good weather 50 percent of the spray above agricultural fields does not reach the target crop. Accidental drift of endosulfan, which is very toxic to fish, can entirely eliminate fish from streams, rivers, and ponds.

In Canada large tracts of coniferous forests were sprayed with DDT to control larvae of the gypsy moth. The rivers running through the forests contained after the spraying sufficient DDT to kill off most aquatic invertebrates, which in turn caused most of the fish to die by starvation.

In Madagascar aerial sprayings with pesticides have had extremely adverse effects on the whole fauna. Such regular sprayings transformed for months vast areas into real deserts void of animal life, but dead birds, reptiles, and amphibians "covered the soil" (Arnoult, 1970).

In other parts of Africa, particularly Senegal, Mali, Niger, Chad, Nigeria, and Cameroon, a highly toxic pesticide, parathion, is used to destroy seed-eating birds of the genus *Quelea*. Parathion is chiefly used by night in these birds' roosting sites, which are often located in or near aquatic habitats, for example, in reed beds. Such operations have had disastrous consequences for many other bird species which often in thousands occupy the same habitats for sleeping, such as cormorants, herons, egrets, swallows, yellow wagtails (*Motacilla flava*), and others. In addition, ducks, waders, rails, gallinules, and certain warblers frequent these habitats permanently. Moreover, all the fish, amphibians, reptiles, birds, and mammals killed by parathion attract storks, herons, kites, harriers, and eagles which accumulate in the area and in their turn get poisoned. When parathion is sprayed by air it even reaches marginal areas and contaminates the living organisms there.

In Cameroon thousands of aquatic birds were killed in 1970 as a result of the use of pesticides against queleas. In Senegal parathion killed hundreds of herons and also warthogs.

These figures are only minima and the records accidental, because after each treatment with parathion the area concerned is dangerous to visit and, therefore, banned for humans for thirty-six to forty-eight hours.

Aerial spraying with toxic chemicals over vast areas of temperate regions of Europe, Asia, and North America to get rid of undesirable trees like birch and aspen in order to favour monocultures of conifers is an eradication of vegetation which, beside its side effects on animal organisms, has many other adverse effects as discussed in the chapter on vegetation.

Chemical defoliation of trees for military purposes, which for years has been going on in the rain and mangrove forests of Vietnam, does not, in general, seem to virtually defoliate the trees but rather to change the color of the leaves. They become reddish-brown but usually remain in the canopy for long periods. The ecological effect of this drastic environmental change is still unknown, but obviously the impact on later vegetation (other than the "defoliated" trees themselves) and animals must be profound, particularly since such large areas are involved. The long-range effects on the ecosystem seem so far to be entirely unknown but certainly constitute hazards to living organisms.

Recently (January, 1971) the results of an investigation on the use of herbicides to defoliate large forest areas, carried out in South Victnam by the American Association for the Advancement of Science, show that about one-seventh of the rain forests of Vietnam has been sprayed with defoliants since defoliation methods were first be used in 1963. The effect in some areas is so drastic that the regions concerned appear to have been exposed to bombing with nuclear weapons. One-fifth of South Vietnam's 14 million acres of commercial hardwood forests, worth an estimated $500 million, has been devastated. Moreover, about half of Vietnam's mangrove forests, mainly in the Mekong Delta region, had been completely destroyed with no sign of regeneration after three or more years. This destruction will certainly lead to coastal erosion and reduction of a major source of fuel wood and charcoal, to say nothing of the other effects due to the collapse of an aquatic-terrestrial ecosystem with its many interrelated bio-communities. Other major targets for defoliation spraying were mountain rain forests which had been transformed to bare lands.

A few days before the release of the defoliation report, the United States Government announced that it was "initiating a program for an orderly yet rapid phase-out of the herbicide operations" in Vietnam.

Another report from Vietnam by Nguyen-Van-Hiep, Chief of the Service for Conservation of Nature of South Vietnam, was published by IUCN in 1971. The defoliants have destroyed hundreds of thousands of hectares of rain forests and caused changes in the climate and an invasion of bamboos and shrubs which may make it impossible to restore the forest unless expensive re-planting can be achieved in time. About seventy species of trees have been damaged by the defoliants, of which not less than thirty-nine species have died. Six species did not show any influence of the treatment. This report is shocking. Never in the history of man has such a tremendous forest destruction occurred in so short a time. It is for the moment impossible to estimate the losses suffered by the fauna, but the report indicates that many mammals, birds of prey, and aquatic birds in particular have decreased.

In Sweden, a country where the use of mercury compounds for seed-dressing has reached quantitative proportions far beyond what was necessary for agriculture, mass deaths have occurred among birds. Thousands of rooks (*Corvus frugilegus*) have died from seed-dressing poisoning. Other seed-eating passerines like the ortolan bunting (*Emberiza hortulana*) and the yellow bunting (*E. citrinella*) were locally exterminated over wide areas of southern and central Sweden, while populations of the same species in northern Sweden, where agriculture is less developed, maintained their frequency levels. In 1958 at least 10,000 migrating starlings (*Sturnus vulgaris*) were killed in one area alone by feeding on insects contaminated by DDT and other herbicides.

As in Great Britain, the number of birds of prey and owls in Sweden has greatly decreased due to secondary poisoning from eating contaminated prey. The white-tailed eagle (*Haliaeetus albicilla*), peregrine (*Falco peregrinus*), kestrel (*F. tinnunculus*), goshawk (*Accipiter gentilis*), sparrow hawk (*A. nisus*) and eagle owl (*Bubo bubo*) have all been seriously affected. Analyses have shown high residues of mercury compounds both in the bodies and in unhatched eggs. The white-tailed eagles breeding in the archipelagos of the Baltic Sea have almost failed to reproduce, and their numbers have fallen off to a dangerously low level. In fact, the Baltic Sea has been so polluted by DDT and mercury compounds that it is dangerous to eat some Baltic species of fish. The white-tailed eagle feeds on fish, wounded or sick animals, and carrion. Of all birds in the Baltic area it shows the highest concentrations of poisonous compounds—an average of 25 milligrams per kilogram of fat. This has resulted in failure of reproduction, so the species is now the most endangered bird of Sweden. Baltic seals, which feed entirely on fish, contain 125 milligrams per kilogram of fat. Salmon average 31 milligrams per kilogram. Concentrations of this level are unknown elsewhere.

In the United States the story is much the same for the peregrine, the bald eagle (*Haliaeetus leucocephalus*), the osprey (*Pandion haliaëtus*), and a number of other raptors. Before North American birds of prey were affected by DDT about 11,000 hawks were seen on migration at Hawk Mountain in Pennsylvania in a single day. At present the number is less than half. Sea and freshwater birds are also marked victims of toxic chemicals, because they too are at the end of the food chains. Many aquatic birds like grebes, pelicans, egrets, herons, and gulls have died in North America from pesticide poisoning. Particularly, the brown pelican (*Pelecanus occidentalis*) is decreasing rapidly. It has been pushed to the brink of extinction over large areas of its range in North America, both in the Gulf of Mexico and in California. For example, of the pelican colonies on Anacapra Island off California, only four pairs out of 550 breeding pairs laid hatchable eggs in 1970 due to thin-shelled eggs. It was the same for the cormorants, also fish-eating birds.

Many similar instances could be cited from many European countries. Birds dominate in the reports, but terrestrial and aquatic —even marine—animals representing all classes of vertebrates have been found to contain pesticide residues. So has man. Of all vertebrate animals under threat from toxic chemicals, birds of prey and owls are almost certainly the most seriously menaced. Locally some species have been all but exterminated. Of thirty-seven species of diurnal raptors breeding in Europe, the populations of twenty-five have decreased in recent decades, many of them drastically, due to pesticides. The disaster has been greatest in those areas where pesticides have been most heavily used.

A study in Sweden on the increase in the mercury contents of feathers from raptorial birds (based on skins stored in museums) over the past decades by comparison with the contents observed prior to 1940 showed that the increase was due to the use, in agriculture, of insoluble compounds of mercury. After the ban in Sweden on mercury as cereal seed dressings, the percentage of this compound in feathers of birds of prey was brought down to its former, normal amount.

There are many ways in which organo-chlorines affect birds. The transmission of dieldrin, for example, from dressed wheat through bank voles and field mice to kestrels and tawny owls has been studied by the Nature Conservancy in Great Britain. One kestrel died with a liver residue of 11.9 ppm dieldrin after feeding on the equivalent of seven to eight voles in six days (Cadbury, 1970). Changes attributable to pesticides in egg breakage frequency and eggshell thickness (decrease in weight) in the peregrine, the sparrow hawk, and the golden eagle in Great Britain from 1951 onwards, combined with direct poisoning, led to an unprecedented breeding failure and subsequent population decrease on a national scale for the peregrine and the sparrow hawk. The same phenomenon of eggshell changes were shown by a number of bird species during the same period. The only known environmental change which parallels the eggshell change in timing and geographical pattern, and has a likely physiological connection, is the widespread contamination of the ecosystem by residues of synthetic organic chemicals used as pesticides (Ratcliffe, 1958, 1960, 1967, 1970).

In the United States similar findings have been recorded in buzzards, eagles, and ospreys. The eggshell thinning in birds, caused by DDT poisoning, is found not only in birds of prey but also in pelicans, cormorants, common egrets, ibises, ducks, gulls, and murres. It has contributed to local crashes in several bird species, and there is now evidence from the field and laboratories of the biochemical pathway between DDT and the eggshell thinning, leading to death, embryological defects, reproductive impairment, behavioural alteration, physiological collapse, and changes in hormone levels—all probably symptoms of the same fatal process. Also biphenyls cause eggshell thinning in birds.

Globally, eggshell thinning in birds has already been found in at least eight orders and forty species of raptorial and fish-eating birds. DDE, the most persistent DDT compound, has been strongly associated with thc eggshell thinning process. Dieldrin has killed birds as large as eagles.

Amphibians and reptiles are insect-eaters and many species are undoubtedly reduced by pesticides to such an extent that they are locally exterminated, although this has not been investigated so thoroughly as in birds and fishes. Many frogs and toads are victimized. In tropical Africa, for example, tree frogs (*Hyperolius*) have disappeared or decreased enormously in agricultural regions, where they formerly were very common. Where their musical voices formerly filled the night, there is now complete silence. Many species of lizards have vanished. Man has killed his allies in the battle against insect pests.

Mammals are affected. In Europe the otter, feeding on fishes, has almost disappeared except in wild areas. In the Waddenzee off the Netherlands the number of common seals has dropped from 1,450 in 1968 to 950 in 1970. Many seals have been found dead. Not less than 22 percent of the bodies collected in 1970 showed a death by mercury poisoning. Bats are much more sensitive to DDT poisoning than are other small mammals. In Arizona the populations of guano bats (*Tadarida brasiliensis*) have decreased drastically. In 1963, one colony exceeded 25 million individuals. In 1969, the total population was estimated at about 30,000 bats. The cause of this reduction is not known, but guano bats travel miles each evening to feed over irrigated agricultural fields and they migrate seasonally to agriculturally developed areas in Mexico. In 1968, the hillside outside a colony of guano bats was "covered" by dead and dying bats. Those tested were negative for rabies but contained DDT and DDE in their body fat. The tremendous decrease of guano bats at only one colony in Arizona means that these bats no longer destroy, as they once did, 40 tons of insects nightly (Cockrun, 1969). Thus, insecticides have apparently eliminated one of the most efficient natural insect killers from the area.

The claim that agricultural treatment with pesticides is followed by higher productivity and crop quality has been repeated so often that its accuracy is taken for granted. But when neutral scientists have looked into the matter, they have often found a different picture. One therefore questions the general validity of the pro-pesticide statements. Very little seems to be known about how to determine the quantities of pesticides required to cope successfully with the parasites without causing undesirable side effects.

Apparently a sufficient accumulation of pesticides in vertebrates has a detrimental effect on reproductive capacity. This is an indication of what might happen to man. We know nothing about the long-term genetical effects of pesticides on vertebrate populations. In employing toxic chemicals in our time man is playing with a hazard that may affect generations not yet born. We know that in a number of mammalian species, including man, DDT passes from mother to offspring through the placenta. Breast-fed human infants in various parts of the world are at present getting twice as much DDT as the World Health Organization's recommended daily minimum intake. When the same amount of DDT has been given in the food to laboratory animals, they begin to show biochemical changes. The biological effects are undoubtedly dangerous.

To summarize the essential facts about pesticides, it is certain that several persistent biocides are deadly environmental concomitants to which higher vertebrates also fall victim. But they are more than that, because they poison the ecosystems by passing from organism to organism. Hard pesticides such as DDT, aldrin, dieldrin, endrin, lindane, heptachlor, and chlordane do not disintegrate for fifteen years. Even small amounts can cause death, reproductive failure, and behavioural abnormalities.

Yet it is predicted that the use of pesticides will increase more than twofold by 1975, and there will be great expansion in the use of herbicides. Obviously, as the use of pesticides continues to increase quantitatively and geographically, the contamination of the environment will also increase.

The subject of pesticides and its implications for animal populations is so complex that it cannot be dealt with thoroughly in this section. However, only one conclusion is possible: the blind destruction cannot be permitted to to go on indefinitely. The present ruthless use of pesticides must be stopped. All such highly destructive pesticides as DDT, dieldrin, aldrin, heptachlor, endrin, lindane, chlordane, mercury, and other persistent chemical compounds used to kill organisms should be completely outlawed. The use of other pesticides that are not "hard" should be permitted only after careful ecological and medical considerations, as a last resort tool when other methods fail. Manufacturers of chemical pesticides should be encouraged by governments to develop effective, nonpersistent pesticides harmless to nontarget organisms. If this is impossible, which is difficult to believe, it will be necessary to use biological controls to replace pesticides gradually.

The use of pathogens like viruses, bacteria, and fungi has been successful as a pest control in some cases but represents risks to human health. Selective breeding of resistant species of plants is another means to avoid pests.

Biological Control

Is there an alternative to the pesticides? Should proposals for all other forms of control be dismissed? Before discussing this question, it is important to state that not all chemical pesticides or fertilizers are harmful. But many pesticides are applied indiscriminately, and often in unnecessarily large quantities. Further, the manufacturers encourage routine applications of pesticides in large quantities without concern for intelligent use of these compounds or for other forms of control measures. The tremendous commercial interest in the use of pesticides has led to neglect of the possibilities for using other control means. Potential natural enemies of pest species, disease-resistant crop varieties, crop rotation, diversified environments around and between crops, are some environmental means to use in the battle against pest species in monocultures.

Those who advocate persistent application of toxic chemicals in forestry and agriculture often rule out biological control. They claim that biological control has proved a winner in only a few instances. This is a comfortable attitude, but unwise and dangerous. Moreover, the claim is false.

As long as 800 years ago ants (*Oecophylla smaragdina*) were reared and used in China to protect the orchards from pest insects. The same method was also practised in the East Indies. In Germany ants of the genus *Formica* are colonized in artificial nests and used to protect forests from pest attacks.

Systematic large-scale methods of biological control have been used since the late 1880's and have proven highly successful. A review based on DeBach (1964) and Simmonds (1965) cites sixty-six instances where complete control has been attained, eighty-eight showing substantial control, and seventy-one partial control.

The California citrus industry, threatened with virtual destruction by the cottony cushion scale (*Icerya purchasi*), was saved in 1888–89 by the introduction from Australia of the ladybird (*Rodolia* (*Vedalia*) *cardinalis*), a beetle. This is one of the earliest biological control successes, and *Rodolia* has been successfully introduced subsequently into many areas to control *Icerya*. The citrus black-fly (*Aleurocanthus woglumi*) was very successfully controlled in Cuba and Jamaica by introduction of the parasitic wasp (*Eretmocerus serius*) from Malaya (1930). Woolly-aphis of apples (*Erisoma lanigerum*) has been controlled in many areas, including Europe, Australia and New Zealand, following the introduction from America of a parasite, *Aphelinus mali*. In the island of Principe off West Africa, the accidental introduction of the coconut scale (*Aspidiotus destructor*) in 1952 caused a 70 percent reduction of the copra crop by 1958. Completely successful biological control was achieved by introduction from Trinidad of a ladybird, *Cryptognatha nodiceps,* and the copra crop was restored to normal. For a total expenditure of about £2,000 an annual loss of some £100,000—of considerable importance in the economy of a small island—was eliminated. These are only a few of many examples that could be cited to show highly successful biological control of insect pests.

Very effective control of the Giant African snail (*Achatina fulica*) has been obtained in Hawaii and several Pacific islands by introducing predacious snails. In the field of weed pests, the spectacular biological control of *Opuntia* cactus in Queensland over an area of some 70 million acres, and also in Hawaii and elsewhere, is a classic example, as in a smaller way have been the controls of *Cordia macrostachya* in Mauritius and *Eupatorium adenophorum* in Hawaii, New Zealand, and to a lesser extent in Queensland. The story of *Opuntia* is worth telling in some detail. This prickly pear cactus was introduced into Australia from Mexico for livestock feed. It did not take long before the cactus overran millions of acres, changing natural rangeland to dense, impenetrable thickets. All attempts to control this plant pest failed, until a moth (*Cactoblastis*) was brought to Australia from Argentina in 1928 and 1930. This insect feeds on *Opuntia*. It reduced the cactus, and by 1933 the last big area of prickly pear in Australia had been cleared out (Graham, 1944).

In these biological case histories control is permanent and does not require recurrent annual expenditure. It is quite true that many efforts at biological control are either completely unsuccessful or partially so; but the economic benefit achieved has been far in excess of the research funds employed in all biological control work. And the total of these funds is very small compared with the amounts expended in developing and testing new chemical insecticides before they are put on the market (Simmonds, 1965). The costs of discovering a chemical and developing it to the stage of practical use has been put at anything up to one million pounds sterling, according to the Shell International Chemical Company in Great Britain.

The biological control of pest insects on the Fiji Islands is another classical example of the usefulness of cooperating with carefully selected natural predators. Several insects belonging to quite different groups have been successfully controlled or entirely eliminated by natural predators introduced without causing negative side effects (Taylor, 1937). Other instances of successful control are those of the eucalyptus snout beetle (*Gonipterus scutellatus*) in South Africa by a parasitic wasp (*Anaphoidea nitens*) from Australia, and of the sugar cane leafhopper (*Perkinsiellia saccharicida*) in Hawaii by a predacious bug (*Cytorhinus mundulus*) from Fiji. Thynnid wasps caught in Chile and Argentina have been introduced to the grass grub in New Zealand.

One of the most successful methods of biological control is sterilization by irradiation of males of insect pest species. The classical example is the experimental control of the screwworm (*Chrysomyia macellaria*), a fly on the island of Curaçao. This island was completely freed from this pest in the preliminary experiment. Later the real control of the screwworm using the sterile male technique took part over large areas of Florida and Texas, where virtual elimination of this fly has been achieved. The same method has also been used successfully against fruit flies in some Pacific islands. Although the sterilization method cannot be easily applicable to all pest insects, it has a great potential and should be used whenever possible.

Another field of biological control is the use of insect pathogens or specifically bred insects to inject lethal genes into pest populations. Vertebrates also are used. The mosquito fish (*Gambusia affinis*) has been utilized in many parts of the world for the specific purpose of controlling mosquitos by consuming the larvae. The grass carp (*Ctenopharyngoden idellus*) is regarded as a means for controlling unwanted plants.

Controlling pest species by using other living organisms that prey on or parasitize them is the simplest and usually the most rapid biological method. Usually this technique is highly selective during the pest explosion, but does not necessarily turn out badly when the pest species is eliminated and the predator remains on the scene with established populations. The case histories from Fiji, in particular, show convincingly how effective this method is even in the long-term perspective. However, it is evident that man's cooperation with and selection of predatory species should be sought among native or established beneficial organisms. In general, native pest species are controlled by native predators. But when a pest has been introduced to a new area, where it can proliferate because its natural parasites and predators are not present, the latter should be imported after careful selection, and not before field experiments in smaller, isolated areas have proven that no adverse environmental effects will follow.

Although the principle of providing a diversified, optimum environment is a sound way to get aid from natural enemies of pests when such assistance is needed, the natural predators are not always able to keep down pest outbreaks. When DDT entered the scene, the use of biological control was largely ruled out as a method because DDT was so cheap and relatively easy to use. DDT in large quantities was even applied to control U. S. citrus pests other than the cottony cushion scale, which for fifty years had been effectively controlled biologically by a beetle (*Rodolia cardinalis*). This addition of a chemical control to a biological control gave an instructive lesson. The DDT had no effect on the scale, but killed its natural predator, the *Rodolia* beetles. In two years of DDT applications the scale increased severely, while its control predator disappeared. The latter had to be reintroduced, and the DDT application abandoned.

As pesticides and biological control are used chiefly in areas with high agricultural or forestry development, it is stimulative of thought to look back in time and consider the productivity of areas known from history to have been highly fertile during hundreds or thousands of years. There are many such areas in the Mediterranean region for example. Sicily, southern Italy, and parts of Spain were continuously important granaries during the Greek and Roman eras. These lands were intensively farmed for millennia and produced up to four or five crops annually without losing fertility. The cultivated fields were located in a biologically diversified country with woods, copses, gallery forests, and wetlands rich in plants and animals. Today, the same areas have lost these environmental benefits through overexploitation, overgrazing, overtrampling, overburning. Despite heavy use of pesticides, these areas approach a state of total dereliction. Biologically they have collapsed.

This decline of previously fertile lands emphasizes the importance of developing long-term research programs on biological controls, of which the environmental diversity is just one tool. If only a fraction of the sums used for research in the field of environmentally dangerous, generally destructive, and increasingly more toxic chemicals could be channelled into biological control research, with its promise of future environmental recovery and human survival, it would undoubtedly be a great service to mankind.

Population Explosions and Crashes

When speaking of population explosions, most people automatically—and rightly—think of human beings. The human population is, indeed, involved in a series of chain explosions, and this is basically the most serious conservation problem of our time. Sudden population highs in animals are often of environmental significance without being detrimental.

Factors governing fluctuations in animal populations have been mentioned briefly in previous sections. Voles and related rodents show periodic fluctuations, fairly often including spectacular highs, which usually are followed by crashes—i.e., sudden decreases. Such phenomena may also occur in other herbivorous mammals, as well as in birds and fish. They may result in wanderings by the burgeoning species over vast expanses. The mechanisms behind these population oscillations are still partially unknown, but apparently several factors are involved. The advantages of sudden crashes for the environment—escaping vegetation destruction—and for the species—survival value—are obvious.

This section is not intended to deal with the complex population-regulating factors and mechanisms, which involve physiology, behaviour, ecology, and evolution. Its purpose is simply to illustrate how tremendously productive various habitats are and how rationally nature avoids catastrophes that might disrupt the economy of a biocommunity. This is a conservation lesson to man, who artificially eliminates the natural barriers preventing the catastrophe caused by population excess. However, many of these natural barriers are invisible and in the long run cannot be avoided.

Phenomena associated with population density and explosions are apparently connected with endocrine responses. Stress symptoms in overcrowded populations of rodents show similar features and effects to those exhibited by men living in densely populated cities. According to Christian (1957, 1959), in the case of rodents the "social pressure is the fundamental factor limiting population growth." This is in line with the data obtained by the present author while working on the Norway lemming (Curry-Lindahl, 1961, 1962, 1963). In this species we find that neither predation nor food shortage plays any decisive role in the population crash. The Norway lemming is probably unparalleled among mammals in showing spectacular population highs followed by drastic declines. However, the social pressure in stressful conditions releases metabolic and endocrine derangements leading to physiological collapse, with mass death in its wake.

This dramatic event is just a part of long-term interrelations between a species and its environment. This complicated system has been explained by Christian (1961) in a wider environmental context, physiologically and socially, in the following way:

> It seems that a complex system of feed-backs which respond to socio-psychological pressures have evolved which regulate and control population growth and keep the population within environmental limits but not without pronounced fluctuations. These feed-backs manifest their effects largely through the reproductive system and secondarily by increasing mortality. Selection must operate largely through the dominant individuals in a population, as in most cases these bear the major load of reproduction and, furthermore, their young are the most likely to survive and reproduce normally. The growing evidence for the profound prolonged effects of increased density on offspring born during periods of increased density has important implications on the selection process as well as for the growth of populations, especially in determining the intervals between major peaks of density. The fact that these physiological reactions are brought about largely by behavioral phenomena represents progress in the evolution of independence from environmental fluctuations, as contrasted to direct dependence on food, climatic changes, and other factors of the environment, for the regulation of reproduction and numbers by eliciting directly these physiological responses. Even the annual cycles of reproduction and numbers of Microtus are largely density-dependent. Therefore mammalian populations, at least of the species so far studied sufficiently (Mus, Rattus, Microtus, Clethrionomys, Marmota, Oryctolagus, Lepus, Sylvilagus, Cervus and Myocastor) regulate the sizes of their populations largely through internal behavioral-physiological interactions. Insofar as experiments are analogous and permit conclusions, dogs, guinea pigs, monkeys, and man respond similarly to increased numbers, at least in terms of increased secretion of adrenocortical steroids. When environmental factors do exert controlling effects, they probably do so largely by altering the social or competitive situation and thereby shifting social pressures up or down, rather than

> by acting directly. The evolution of the social regulation of population growth could be considered a marked developmental advance over direct environmental regulation, and coincides with the greater development and importance of the higher central nervous system and warm-bloodedness in mammals in contrast to the lower animals. Finally, since the survival and evolution of a species must depend on reproduction by dominant individuals, the selective process would seem to be operating in a direction to increase the importance of behavioral adaptation.

As the above indicates, Christian regards the social pressure as not only the most important limiting factor in populations of mammals but also the fundamental population regulating factor. This assumption does not, of course, necessarily imply that social pressure is always the decisive limiting factor. As far as regulating factors are concerned, there are always several other environmental ones in operation within a population. Many of these factors are probably of the same or even higher importance for the regulation of numbers as is the social pressure factor. The latter is chiefly, or only, operative in extremely dense populations and, therefore, becomes a limiting factor at final stages.

Rodents provide the most well-known examples of self-regulatory mechanisms in the control of animal populations. It has been suggested (Klein, 1968, 1970) that in the case of ungulates there appears to be a relationship between the self-regulatory ability of their populations and the relative stability of the environment within which they have evolved. North American deer that are adapted to early successional stages of vegetation, which are of transitory nature, do not appear to have well-developed self-regulatory mechanisms and are normally characterized by wide population fluctuations. On the other hand, the roe deer (*Capreolus capreolus*) in Europe and bovids such as the Uganda kob (*Kobus kob*) in tropical Africa, and the North American mountain sheep (*Ovis canadensis*) that are found on relatively stable vegetation types, appear to have evolved behavioural mechanisms that tend to contribute to the stability of their populations. However, in the latter case predation may play a role, as it certainly does for deer populations in Alaska and certain North American islands, where wolves tend to "level out" fluctuations in numbers of deer in contrast to situations where wolves are not present.

Presumably mechanisms for population regulation have evolved in close relation to the degree of environmental stability. Therefore such mechanisms must be expected to differ widely in various ecosystems in response to the degree of environmental stability or instability.

In this connection the importance of behavioural mechanisms in the regulation of animal populations in relation to their food supply must be stressed. This factor has been particularly elucidated in a recent review by Watson and Moss (1970).

The population dynamics of vertebrate animals can teach man a lot. More than ever before we are faced with the challenge of adopting a biological approach to man's population crisis.

Introductions of Exotic Species

By "exotic" is meant a species that is not native to the environment, or the area where it lives or is going to be introduced. The biological effects on the environment produced by the introduction of exotic animals have often been considerable. In some cases whole ecosystems have collapsed; in others, habitats have changed completely. Many such introductions have been economic disasters. Others have had indifferent results. A third category has paid off economically. In several parts of the world introductions of exotic game animals have been successful.

Often exotic subspecies or populations are introduced to areas where the same species occurs in natural, long-established populations that are well adapted to the local climate and other physical and biotic conditions. These environmental factors have had a positive impact on the local population which has developed the best genetic equipment for coping with that environment. Introducing exotics may result in lowering the quality of the local population either through competition, breeding, or exposure to exotic parasites or diseases to which it is not resistant.

Many of these introductions are not licensed, or even recorded. Scientifically this is regrettable because introductions of exotic species, subspecies, and populations—particularly of fish—create a zoogeographic, genetic, and taxonomic hodgepodge that creates obstacles for future research.

But the adverse ecological and economic implications of exotic introductions may be much more serious than that. In too many cases the introduced species has increased in number and become a pest destructive to the environment and impossible to get rid of. Useful plants and animals have been reduced or wiped out. The detrimental effects can undermine entirely the productivity of an area, alter ecosystems, provoke erosion, drought, and famine.

Introductions of exotic species, or transplantings of fish from one water system to another, have been made from early periods of human activities. However, it is only within the last two hundred years that intercontinental introductions have been made deliberately, some of them repeatedly and on a large scale. The majority of these attempts have failed. Many of those that "succeeded" are the real failures because they have caused tremendous economic losses to man. Some have been catastrophic. The costs cannot be measured because the values lost are incalculable.

One can never foresee in detail what the ecological consequences of introducing an exotic species will be. Even if an introduction initially appears to be a success, it can later turn out to be detrimental. The long-term ecological effects of a new component in a habitat are almost always too unpredictable to justify any introduction.

Astronomic economic losses have resulted from introduction of exotic species in almost all continents. The best-known example of such a disaster, destroying human food resources as well as rangelands for domestic animals, is the introduction of the European rabbit in Australia and New Zealand. Consideration of this single case history is to make one realize what enormous biological and economic risks are taken when foreign animals are introduced into a country. The rabbit was introduced in Australia in 1859. Three years later the Australians regarded the animal as a national catastrophe. For a hundred years countermeasures were tried, such as the introduction of other animals to kill off the rabbits, but in vain. Instead, these newly introduced animals exterminated valuable native species. The staggering increase in the number of rabbits in Australia is reflected in the statistics on the export of skins: 33,000 in 1873, 9 million in 1882, 17½ million in 1945. In 1950, an imported viral disease, myxomatosis, proved at first to be an effective destroyer of rabbits, but after six or seven years the rabbits became partly immune, and now the number of survivors is greater after each outbreak of the disease.

New Zealand has likewise been the scene of giant failures that followed introduction of numerous exotic species. Not only were pigs, goats, and other domestic animals brought in to provide meat but also wild mammals, such as deer and rabbits, for hunting and sport. One mistake led to another. Several species soon began destroying grain, forest, and native animals. Not less than fifty-three exotic mammals have been introduced, of which thirty-four are still there. When the rabbits, introduced before 1838, became too abundant, predators such as ermines, polecats, and weasels were brought in. They quickly established themselves and, like rats and dogs, turned on the native birds. The few reptiles and amphibians were also affected. Herds of sheep have changed the islands beyond recognition.

The red deer, introduced in 1851, soon made itself at home and is now found almost everywhere. It multiplied so rapidly and did so much damage that drastic control steps had to be taken. In 1931 deer protection was ended, and bounties were paid for their destruction. From 1932 to 1954, professional hunters engaged by the government killed more than half a million red deer. The total number shot during that period is said to have been more than two million. Finally, in 1958 a poison campaign was undertaken. But the deer are still there.

Another serious mistake was introduction of the brush-tail opossum (*Trichosurus vulpecula*) into New Zealand in 1858 as a valuable fur-bearing animal. In New Zealand, unlike Australia, this opossum, which is herbivorous, caused tremendous destruction in forests and orchards. It increased rapidly, and soon the New Zealanders declared a war of extermination against it. In 1945 alone, they killed 922,088 of these animals. To the wild animals must be added domestic animals such as cattle, goats, and sheep. About six million head of cattle and forty million sheep are now grazing in New Zealand. Practically all strata of vegetation from the grasses to the crowns of trees have been seriously damaged, which has led to a further reduction of forest land area. The ground layer has been grazed off, in many places preparing the way for serious erosion.

The Caribbean Islands have notably been the victims of man's mania for introducing exotic animals. Tremendous damage to the native animals has been done and is still being done by these aliens. Introduced dogs, rats, pigs, monkeys, cats, and opossums prey extensively on the indigenous species. Introduced goats destroy the vegetation. Worst of all has been the introduction of the mongoose (*Herpestes auropunctatus*) to many islands of the West Indies. It has led to complete or nearly complete extermination of numerous native mammals and birds. Other island groups like Hawaii and Galapagos have likewise been ravaged by the liberation of exotic animals.

In the United States probably about 80 percent of the pestiferous insects have been introduced—the majority, of course, unintentionally.

The introduction of exotics often releases detrimental chain reactions in the environment. Exotic animals are also unpredictable in the sense that in a new environment they easily develop food habits other than those of their native habitat, as happened in the case of the brush-tail opossum in New Zealand. The raccoon-dog (*Nyctereutes procyonoides*) fed on fish and crabs in its homelands of eastern Asia, but in the Caucasus it turned to hares, game birds, and poultry.

Introduced vertebrate species that have become pests due to damage of human food resources include, in addition to those already mentioned, the European hare (*Lepus europaeus*) in Canada and Argentina, the gray squirrel (*Sciurus carolinensis*) in South Africa, starlings and house sparrows in North America.

Islands are in general vulnerable to the effects of exotic introductions due to the local specialization of both plants and animals, which in many cases have evolved in the absence of mammals. The vegetation therefore cannot resist grazing or browsing, and the animals cannot withstand mammalian predators. Depletion of the plant cover, erosion, and general decline soon follow.

Industrial introductions of exotic animals—chiefly fur animals—have frequently been made in the U.S.S.R. In 1953 alone, 21,000 animals of various species were released in the U.S.S.R. During the period 1927–53 not less than 117,000 muskrats, originating in North America, were introduced in five hundred localities throughout the U.S.S.R. Sportsmen are prone to introduce foreign species that might better withstand the shooting pressure than the native animals. Before World War II Hungary exported 50,000 partridges, 35,000 pheasants, and 40,000 European hares in order to "ameliorate" the game populations in various countries. At present, Czechoslovakia has taken over the same role, exporting not only small game but also larger species, such as red deer and roe deer.

The majority of the crisscross introductions of game species from one country or continent to another for the purpose of recreational hunting enjoyment fortunately has not resulted in establishing populations that could have eliminated indigenous species.

However, there have been instances where benefits resulted from the naturalization of exotic animals, but for domestic animals they have been remarkably few. Unexpected side effects may still occur at any time. Fish dominate among successful introductions, but biologically they are often failures because they compete with, and in many cases exterminate, native species or destroy habitats. For fish this is the case in all continents. The carp is an outstanding negative example in North America, South Africa, and Australia. On the other hand, introduced fish, if carefully selected, often contribute a greater protein output per unit area than many indigenous species.

Introductions of fur bearers, notably in the U.S.S.R., have been economically successful. In that country acclimatizations of animals regarded as economically useful to man have been made on an industrial scale. It is estimated that introduction of exotic rodents in the U.S.S.R. caused an enrichment of the ecosystem, resulting in an improvement of habitats for many local species. Introduction and multiplication of the muskrat produced a reliable and abundant food supply for a number of carnivores. But the reputation of the muskrat in Europe is not so positive elsewhere. The most rapidly spreading introduced mammal in Europe, it is regarded as a pest in many countries due to its serious damage to river banks.

Introduction of exotic species should be avoided or when highly desirable undertaken with utmost care. Thousands of case histories indicate caution. Ecologically, and therefore also economically, it is wiser to manage and encourage already existing species within a region or an ecosystem rather than to try to introduce an exotic species.

Reintroductions

Introduction of a species that once occurred or still exists in an area has, in general, less serious ecological implications than do introductions of exotic species. Reintroduction of a species that has been locally exterminated is often desirable for various reasons. Such reintroduction may serve the purpose of ensuring the survival of a species by enlarging its range. It may also serve to restore a depleted population within a national park or a nature reserve—or even one that has disappeared entirely. In such cases reintroduction is a useful conservation method. Perhaps the best-known example is the successful reintroduction of the square-lipped rhinoceros in many parts of South Africa.

An important principle of reintroduction is that the translocated animals should belong to the same subspecies as the one originally occurring in that area. This is particularly important in cases where there still are individuals of the original population living within the area of reintroduction. Both races are species *in statu nascendi* and should not be mixed. Even within a subspecies, particularly those having a wide range, various populations may have different ecological requirements due to adaptation to local climatic and physical conditions.

Selecting the habitat for reintroduction schemes is also important.

Areas that have lost a species from, say, twenty-five to fifty years before, may have changed ecologically to such an extent that the land is no longer suitable for a species that it previously produced.

The most common motivation for introducing or reintroducing vertebrate species is to provide animals for hunting or fishing. In practical terms, such translations are not difficult. They are almost too easy, which has led to uncritical selection of living material. Also the technique of liberation, and the site chosen for release, have created serious limitations on introductions, because neither the ecology of the species concerned nor the carrying capacity of the environment was considered or understood. This has been the case especially for fish in temperate regions where for decades it has been thought that artificial hatching and introduction of fingerlings were more effective than natural production. Ignorance of the potential of natural spawning and hatching of fish has unnecessarily cost astronomic sums. Introducing individuals of a species into an area that is able to produce the species, in order to increase the shooting or fishing yield, is often bad economy. In general, management of habitats and of wildlife gives better results for less cost.

Tsetse Control and Tsetse Fly Relationships

Probably it is thanks to the tsetse flies of Africa that considerable quantities of large wild mammals still exist in the tropical parts of that continent. About twenty tsetse species occupy parts of tropical Africa that altogether are greater in area than the United States—equivalent to about a quarter of the entire African continent. The presence of these flies on bush and woodland savannas and in forest has been an obstacle to human settlement and to the introduction of livestock on marginal lands. This land has thus been saved from the destruction that usually follows in the wake of grazing and trampling by herds of cattle.

The reason tsetse flies have such an important bearing on land use in tropical Africa is that certain species transmit fatal diseases—the trypanosomiasis—to men and cattle. The human form of trypanosomiasis—sleeping sickness—is not so widespread as the animal disease, called nagana. Both sexes of the tsetse fly feed exclusively upon vertebrate blood. Many mammals, birds, and reptiles in Africa are the hosts of protozoan blood parasites, called trypanosomes (*Trypanosoma*), which do not affect wild animals but may be passed from them to the tsetse when it feeds on them. If a tsetse fly carrying the parasite subsequently feeds on cattle, the trypanosomes are transmitted. Cattle are not immune to nagana. They develop the sickness and usually die within a few weeks. Human sleeping sickness is a different matter because the tsetse fly species transmitting this disease carries it from man to man, while bites from tsetse vectors carrying the animal disease are not fatal to man. However, there is evidence that wild ungulates in exceptional cases can be reservoirs of *Trypanosoma rhodesiense,* which is one of the two trypanosomes pathogenic to man.

The relationships between tsetse flies and their environment, including their mammalian hosts, are complex—as are all interactions in tropical biocommunities. One of the most common tsetses on savannas is *Glossina morsitans.* This species seems to get its most frequent blood meals from the warthog, but it has a wide range of ungulate hosts.

In many areas of Africa, particularly in Rhodesia, an effort has been made to eradicate the tsetse flies by exterminating their hosts. Enormous quantities of wild ungulates—elephants, buffaloes, antelopes, warthogs, bushpigs, and even carnivores—have been slaughtered. These massacres have not led to any positive results, because it is virtually impossible to eliminate from an area all wild mammals that provide food to tsetse flies. Moreover, there is good evidence that at least in some operations, principal tsetse host species have actually increased as a result of expensive control systems (Riney, 1967). Also other unexpected consequences may follow bush clearances. In the Sesi Islands on Lake Victoria, for example, the human population was evacuated because of a sleeping sickness outbreak. As a result, hunting pressure on the sitatunga, a species of antelope, ceased. The antelope increased in such numbers that it destroyed the bush thickets that served as a shelterbelt for the tsetse flies. The tsetse disappeared. Man came back, reclaiming the land. The sitatunga and other mammals retreated, the bush thickets sprang up, and the tsetse flies could breed again.

These flies have special habitat requirements. They are sensitive to direct sunshine during the hottest hours of the day. Therefore they need shade-giving vegetation shelter. Costly thicket clearing to get rid of the vegetation as well as pesticide spraying have been tried, but these attempts have seldom been successful.

Since, as has been shown, marginal lands of Africa are unsuited to cattle raising, it is uneconomic to eradicate wild animals because they are hosts to trypanosomes, and it is uneconomic to clear large areas because they constitute habitats for tsetse flies. Only in areas with established agriculture may it be economically justifiable to reduce the tsetse. Elsewhere a far greater economic return can be obtained from marginal lands by leaving the tsetse there and utilizing the game as a food resource. Although there are many controversial ideas about the tsetse control, most ecologists, conservationists, and even some agricultural researchers now agree that wholesale destruction of large mammals is a wasteful and wrong method to combat trypanosomiasis.

A method of tsetse control much in use at present is selective clearing of tree species which give favourable conditions for the tsetse flies combined with spraying with chemical, persistent insecticides. The side effects of chemical poisoning in the habitats concerned make this method problematic.

On the animal husbandry side it is often claimed that the tsetse is so effective at transmitting nagana that where wild mammals, cattle, and tsetse share a habitat, nagana invariably appears among the cattle. Yet little is actually known of the role wild animals play in the transmission of trypanosomes to cattle. In many areas, the expansion of the tsetse has followed the dispersal of cattle. In the first decade of this century the western Tanzania population of this fly started to expand northward into the savannas of Ankole in Uganda, which were grazed by cattle. By 1958 the whole of the eastern half of Ankole was infested. In this case it was not the wild ungulates that spread nagana to new areas.

The most effective method of tsetse control—or rather trypanosomiasis control—at present is drug treatment. Experimentation in biological control of tsetse flies may develop better methods in the future. In any event, the fact remains that in most tsetse-infected areas of Africa it is better to utilize the existing productive populations of wild animals for meat and hides rather than to eradicate them, the vegetation, and the tsetse flies so as to establish cattle raising, which inevitably leads to further land destruction. There are, in fact, many African problems of greater concern than the tsetse fly and trypanosomiasis.

Value of Marginal Lands for Wildlife Productivity

As the white man's influence swept through the ranges of wild grazing animals all over the world—the prairies in North America, the llanos and pampas in South America, the savannas and high veld of Africa, the steppes of Europe and Asia, the grasslands of arid Australia—indiscriminate land use and tremendous hunting pressure wiped out or greatly reduced incredibly productive natural resources. Dry lands are often marginal lands in the sense that they do not support intense agriculture and pastoralism. In particular, arid grazing lands such as desert shrubs and semidesert grasslands, both in lowland and high plateau areas and regardless of whether they are located on flatlands, rolling plains, or mountain slopes, are highly vulnerable to cultivation and animal husbandry. Preceding chapters have emphasized this aspect.

Through the centuries man has increased the area of marginal lands all over the globe by changing fertile regions to semideserts and deserts. Today most of the terrestrial habitats of the world are in marginal land. But such land has a value providing it is wisely used and managed instead of gradually destroyed. The great value of marginal lands is their productivity in the form of animals. Many of these creatures are large or medium-sized mammals, with a great ability to convert poor vegetation to excellent meat. Many of these mammals do not drink water, or can do without it for long periods. Several examples of this adaptation have been cited. Man should utilize it to his advantage.

The pronghorn antelope in western North America provides an example of how productive wild rangelands once were. Since the first settlers arrived there during the nineteenth century, year-round hunting reduced the population from an estimated level of approximately 40 million to about 30,000 in 1924. Conversely, this same species has responded to conservation and management measures. From 1925 to 1955 the pronghorn recovered phenomenally, to about 400,000 animals (Nelson, 1925, Buechner, 1961).

Man must learn to coexist with natural marginal lands and to restore the man-made ones to their former wealth, measured in terms of productivity. The conservativism of agriculturists, unwilling to recognize that animals other than the few species hitherto domesticated are a potential resource, is distressing. In dealing with crop cultivation and livestock grazing on marginal lands, they always presume that the wildlife are in conflict and must retreat, and that expansion of animal husbandry is always progress. This is not only nonsense; it is a highly dangerous philosophy for man that has already caused the loss of enormous areas of productive lands. It is increasingly important to get rid of unrealistic and conventional use criteria for development of marginal lands. Effective utilization must include that which these areas themselves produce constantly without deterioration: wild vegetation and wild animals for the production of meat, hides, and other products.

For Africa this approach is particularly necessary because with its diverse fauna it has a unique opportunity to develop the living natural resources. Savannas were until recent times the principal food niches both for herbivorous and carnivorous mammals. In most African savanna areas, the great herds of grazers and browsers with their associated predators have been replaced by wandering cattle. This change has not only resulted in lower general productivity, expressed in protein quantities, but also caused fundamental alterations in the vegetation and the soil, the disappearance of surface water, spread of diseases such as trypanosomiasis, rinderpest, and foot and mouth disease—in other words, a lowering of the environmental quality with long-term fatal consequences for man himself.

In 1963 the Food and Agriculture Organization classified the land area of Africa as follows: 8.5 percent cultivated or under permanent crops, 24 percent forest, 19.5 percent permanent pasture, and 48 percent unclassified. As mentioned earlier, about 30 percent of Africa is marginal for any form of agriculture. This portion includes tsetse-infested country, semideserts and deserts that previously produced animals in large quantities. This means that these marginal lands are potentially good pasture areas provided they are used by wild species able to cope with the harsh environment. It is, in principle, the same for other continents. This value of marginal lands in terms of animal products is considerable, and their contribution to the world's protein supply cannot be overestimated.

In planning marginal lands on an ecological basis it is not enough to analyze water, soil, vegetation, and the fauna as separate entities. Although all these elements operate in their own way, they are dependent on each other and form one ecosystem. Fire is also often an important factor. The ecological background is very complex. This is one of the reasons why marginal lands break down when occupied by monospecific combinations of species, e.g., goats, cattle, and men.

The Great Symbiosis: Grasslands and Herbivores

As has perhaps been repeated too often in this book, the grasslands of the world were once the richest protein-producing terrestrial habitats man has ever seen. Yet most of these wildlife areas were in regions of scanty rainfall.

We live in a hungry world, and we cannot afford to leave our crops to be ravaged through destruction of the soil in which they grow. Yet this is just what we have been doing for centuries with wild animal crops all over the world. They have been wasted and plundered to the point where most of the once miraculously productive wild lands are now dead or dying. This, one of the darkest chapters in our history, has provoked an ecological crisis threatening man.

Nowhere has the prosperity of the living earth been so convincingly illustrated to man as in the great plains, wherever they were located in the Old or the New World. Unmatched among our globe's terrestrial habitats, they continuously produced large herbivorous animals in quantities that today appear to be unrealistic. Yet this situation was very recent—in many regions less than fifty or one hundred years ago. For today's generation it may sound incredible that a little more than a century ago a man might have watched by the hour as a herd of bison thundered across the prairie of North America.

Few people of that time fully understood that the pristine landscape, with its tremendous number of large animals, was a dynamic expression of intact habitats that had evolved to ecological perfection and a maximum conversion rate—a kind of giant symbiosis between vegetation and animals. Where the landscape has not been irreversibly undermined by careless habitat destruction, this great symbiosis can be restored. The very idea is a challenge because it offers humanity the possibility of repairing for future generations what it has unnecessarily damaged in the past and present time. The strongest argument for such a scheme is the necessity to counterbalance the factors threatening human survival. Increasing demands for food, as a consequence of increasing human population, cannot be satisfied unless governments and people understand that wild grasslands and wildlife rank among the most important renewable natural resources.

Perhaps the symbiosis spoken of here symbolically is not only a relationship between vegetation and herbivores. In the long run, man might well be involved also, whether he uses wildlife as a source of food or of recreation to relieve the tensions of life.

"Harmful" and "Useful" Species

It is important to avoid classifying animals, and predators in particular, as "harmful" and "useful" species. Such classifications are even used in the game legislation of some nations. The ecology and function of animals in biotic communities are so diversified and complex that they cannot be generalized. Further, the activities of a species changes from one area to another, and from season to season dependent on shifting local conditions. A predatory species should never be doomed or classified as "vermin" before detailed studies have been made of its environmental necessities, its own nutritional value as food for other species, its metabolism, its food consumption in various habitats in the several seasons of the year, its hunting technique and choices of prey, as well as predator-prey relationships, individual specialization, territoriality, population dynamics, cycles, and so on.

Management

Management problems of fishery and wildlife conservation are as varied and as changing as the habitats themselves. They depend on various kinds of land use and on human population pressure. Basically, the role of wildlife management is to keep animal populations optimal, diversified, and harmonized with the environment in order to satisfy the needs of man. These needs are economic, recreational, scientific, educational, social. A sound management policy also requires some basic conservation concepts, including ecological and biological considerations and a respect for native plants and animals, which should always have priority over exotic species. The development and application of management measures to obtain the greatest sustained public benefit from wildlife, or any other natural resource, should never be allowed to go so far as to threaten a species or subspecies with extinction.

Wildlife management includes restoring, protecting, conserving, and maintaining animal populations. All these stages require successful cooperation with the environment, a long-term, ecologically based policy, and synchronization with other kinds of land use within and outside the area involved. No management can ever be successful if it is not based on biological research.

Fish-hatching programs in many countries are based on a management policy of supplying fish for stocking waters either for commercial fisheries or for the enjoyment of anglers. Stocking of fish in artificial waters affected by hydroelectric or irrigation schemes, such as dams, lakes, rivers, and streams, is often made to compensate for the losses of the original fish inhabitants, which vanish due to changes in the water ecology. Some of these species have great economic value and are of international significance because they migrate to the sea, where they become part of the fishery programs of several nations. Examples of such species are several Pacific salmons and the Atlantic salmon. In many temperate river systems fish hatcheries maintain or compensate salmon runs. However, this is a costly method that does not work as efficiently as nature itself. Problems of disease are difficult to cope with in dense populations kept artificially before release.

Many lakes, ponds, and waters are regularly stocked with fish by airplane. This is done on a large scale in many American states to satisfy sportsmen. Unless the stocking program is restricted to species whose natural populations have been reduced by excessive fishing, this form of management has little or no biological or conservation value. It should simply be regarded as a service to the people, sport fishing being the most popular outdoor recreation in North America.

In both fish and game management the habitat is the key factor. The quality of habitats as life zones for animals, including man, is constantly decreasing because man is profoundly changing and destroying the habitats, cover, and food that are vital for the existence of wild animals.

Modern wildlife management must function by planning ahead, foreseeing the future's tremendous human pressure on habitat and wild animals. It can draw a great deal from numerous human mistakes and land misuse during past centuries to reduce as far as possible man's detrimental impact on the environment and wildlife. Much of the damage done to wildlife and its habitats during past centuries is irreparable. Nevertheless management measures should include environmental restorations designed to put back the natural interactions that lead to wildlife fertility in a healthy landscape.

Reserves and Refuges

In wildlife conservation, management and utilization reserves and refuges play an important role. These areas can be highly effective when located and applied properly, but they are not a cure-all for depleted animal populations of all species. Reserves providing breeding grounds, food, shelter, and other vital requirements of wild animals have many values to man. In addition to producing animals that can repopulate adjacent lands suffering from man-made defects, a reserve can also serve as a reference area, giving samples of animal productivity for comparison with areas where various forms of land use depress the animal populations.

It is of great economic and scientific importance for each country or region to have a system of representative habitats, with every type of biome or ecosystem set aside and protected by a reserve. Reserves for larger mammals must be carefully managed unless the area harbours a full set of predators. Without predation, grazing and browsing animals can build up their populations to the point where the available food is exhausted. Such a situation may cause a population crash, and decades or more may be required before the vegetation recovers.

Reserves for migratory species, particularly wildfowl, are an effective conservation tool. Along the flyways from the breeding to the wintering in the temperate regions the hunting pressure is great and the destruction of aquatic habitats is alarmingly widespread. Without chains of refuges, millions of wildfowl would not survive annual migrations. Experiences in the Old and New Worlds show that wildfowl reserves are generally successful. They pay off by increased bird populations.

In the United States in 1960 there were approximately 290 wildlife refuges—in addition to national parks, national forests, state parks, and other nature reserves and sanctuaries—containing 28 million acres. There are 15 big-game refuges with about 14 million acres, protecting bison, pronghorn antelope, Arctic caribou, elk, moose, bighorn sheep, Dall sheep, and grizzly bears. Areas primarily intended for migratory wildfowl include 5½ million acres, and for other migratory birds and wildlife in general about 4 million acres. The U. S. system of reserves has proved to be of great value for the preservation, conservation, management, and utilization of wildlife resources, as have similar systems elsewhere in the world.

The reason reserves are such an efficient conservation and management tool in dealing with animal populations is that they protect the habitat. This is highly significant.

The Animal Retreat

All over the world animals have been exterminated or greatly reduced in number due to man-made actions. The decline of larger species and of commercially valuable species is especially bad. The main causes of the great animal retreat during historic time are a tremendous destruction of habitats and overexploitation. Often habitat destruction and change lead to a heavy decrease in animal populations. Excessive hunting or harvesting gives the *coup de grâce* to a species or brings it to the verge of extinction. Also, species that are thought to be adverse to the interests of man—for instance, predatory mammals and birds—are killed recklessly with the deliberate intention of exterminating them. Trophy hunting, particularly in South America, is also a prime cause of the decline of animals.

In all instances man is responsible for the global animal retreat that has continued for centuries, impoverishing the world. A part of this decline is probably unavoidable due to the increase of the human species. But in most cases the destruction of animals is unnecessary and uneconomic, leading to unbalanced habitats, biocommunities, and ecosystems. In the long run it is against man's interests.

CHAPTER EIGHT

Man

Man, like other animal species, is a part of the environment, of biotic communities, and of ecosystems. Although he modifies his habitats more than any other species and seems to believe that he stands above natural laws, he is still dependent on the ecosystem, like any other organism. This is true for urban populations as well as for those living in the countryside. It is the same for black, red, white, or yellow races of man—wherever they live, in the arctic or in the tropics.

All previous chapters have dealt with the earth's renewable natural resources. Now, we turn to just one species, the only organism in the world that is challenging his own environment—man. The fact that he is doing so in a way far from advantageous to himself leads to the inevitable question: Has his formidable technical development, the product of the evolution of the human brain, outrun his wisdom for using that development in a biologically sound manner?

Ecology of Man

Human ecology is complex, but in principle is the same as that of an animal species. Populations characteristically increase in size until a more or less equilibrium level is reached, around which the population size fluctuates more or less irregularly according to the capacity and variability of the environment. This is a fundamental rule of ecology, which man cannot escape.

However, man gradually—thanks to medical and technological progress—eliminated one environmental barrier after another to such an extent that human populations during the last one hundred years have increased almost unchecked by environmental factors. In the long view this is a highly dangerous situation, both for the environment and ultimately for man himself. It is virtually a death trap because man cannot escape forever from being hit by environmental resistance in the form of unforeseen ecological boomerangs. Such ecological barriers may take the form of diseases, viruses, famine, natural catastrophes, collapses of society, warfare, or other biological manifestations of poor environmental health. Yet we do not know when, where, and with which force the inevitable environmental boomerangs will strike back at human populations. Moreover, most people and governments do not seem to be aware of ecological fundamentals and the existence of limiting environmental factors. The idea of trying to cope with the present rate of global population increase by producing more food is not only a utopia but it also leads to a continuous destruction of natural resources on a monumental scale that will bring us to the point of no return.

All of the major environmental problems that threaten the future of mankind are caused basically by one factor: too many people. Air, water, and soil pollution, overexploitation of renewable natural resources, scarcity of food, poor living standards, lack of harmony, wars, and so on are consequences of a population increase beyond the carrying capacity of the environment.

Tragically, shortsighted overexploitation and plundering of natural resources have also taken place in countries without population problems. In industrial countries the drive to increase the national income and standard of living has led to reckless exploitation without regard for the values of environmental qualities. This combination of blind greediness, ecological illiteracy, and lack of long-term perspective with regard to living nature is man's greatest shortcoming. It will be the basis for an indictment of twentieth-century humanity by tomorrow's generations. People have not yet realized the harmful effects that technology and so-called social welfare have had on the environment.

Man is an ecological paradox in the sense that as a species he exercises a high degree of dominance over other species, but thus far has been incapable of controlling his own population. While animal species are favoured by their total dependence on the natural environment, the human species is disfavoured by the artificial means man employs to increase his numbers above the environmental carrying capacity. The increase will ultimately lead to an inescapable population crash far exceeding the crashes that periodically and naturally hit animal populations.

Vertebrate populations have in varied degree evolved means of adapting to their environment. Many of them are sufficiently flexible to cope with various kinds of environmental changes. They also seem to have developed population control mechanisms to prevent overpopulation, overutilization, and environmental destruction that are fatal to their species and would lead to extinction. Man is biologically equipped with similar means of protecting his own species, but he does not use them, and eventually this must hurt him.

The danger of neglecting ecological realities becomes clearer every day. Yet man seems to assume that his technological ability will solve every environmental problem. The western nations are imprisoned by this philosophy, which they try to transmit through aid projects and development plans offered to undeveloped countries in various parts of the world. The long-term results are a degradation of renewable natural resources at an unparallelled rate. For decades ecologists and conservationists have tried to warn governments and societies of the danger of ignoring the fact of man's dependence on the natural environment. Little or no attention has been paid to these voices. Conservationists were regarded as unrealistic idealists not worth listening to. Since their arguments and statements often were discomfitting to governments and politicians, it was in the latter's interest to exclude ecological expertise from all society and development planning. Mankind is now paying for this serious mistake and will continue to do so for generations.

The thoughtlessness of governments in provoking unnecessary ecological situations that are potentially dangerous to society and human progress originates in their ecological illiteracy, combined with politicians' unfortunate way of not looking ahead beyond the few years of their political mandate. Their inability to think in ecological terms produces a lack of vision. That is a dangerous shortcoming for a biological being who is reshaping the earth.

When in recent years the rapid degradation of renewable natural resources reached proportions that threatened to become a public menace, some governments suddenly woke up and changed their attitude, becoming a bit more positive toward conservation. The question is whether this change in opinion has come too late, or there is still time to repair the damage by restoring the environment through far-reaching global measures based on ecological facts and principles. Without an understanding of man's proper relationships with natural resources, it will never be possible to satisfy human needs—physical, economic, social, and spiritual.

Man's dependence on the natural environment becomes obvious by analyzing the character of his biological stability. Although the range of his biological adaptability is impressively large, permitting him to move into new habitats within a wide ecological range, he cannot escape from the bondage of his evolutionary past. Wherever man goes on the sea, on land, or in the air, he remains a mammal that must retain a microenvironment basically similar to the one in which he evolved. On the other hand, man's remarkable ability to adapt to different habitats, and to tolerate conditions entirely different from those of his original environments, seems to indicate that he can, after all, survive successfully in overcrowded cities and industrial agglomerations, in a world polluted with toxic chemicals, waste products, noise, and excessive stimuli. But the long-term effects of living in such an environment are not fully known. Physical and mental health is certainly affected. Stress symptoms of various forms increase. They are danger signals for population welfare. They indicate the human need for environmental diversity.

Accelerating Evolution

A striking feature in the cultural evolution of man is the time factor. Man appeared on earth about 2 million years ago. The implications for his early cultural evolution were certainly vast, because it was only about 9,000 years ago when man domesticated plants and about 8,000 years ago when he domesticated animals. Man's decisive steps in his cultural evolution have been taken during the last 2,000 years. Of this latter period it is the last hundred years that have seen a tremendous technological evolution with almost unlimited possibilities. This evolution is now accelerating so fast that it creates great risks for the species concerned, because for the first time in man's evolution he is consuming the resource capital of the planet he inhabits. In fact, it is the first time in the history of the earth that one species is depleting the environmental resources and, moreover, doing it on a global scale. Man's accelerating technological evolution without ecological breaks is a deadly danger not only to himself but to the whole living world.

The Population Crisis

It took man about 2 million years to reach a population level of one billion. That was about 1850, but only 80 years later the figure had doubled. In 1970 the world population reached 3.6 billion—almost a redoubling in only 40 years. If the current growth rate, 1.9 percent annually, is maintained, the human population will increase to a figure in excess of 7 billion in just over 30 years. Another estimate is that the optimum world population should be in the vicinity of half a billion, providing the present environmental problems can be reduced to a minimum and be maintained on that level for periods of hundreds of years. Even the figures for areas in the western world located far from the real "explosion regions" in Southeast Asia or Latin America are striking. To give just one example, in New York State the population, 18 million in 1970 (a change from 1969 with +8.4 percent), is expected to rise to 22 million by 1985 and to about 30 million in the year 2000. The population increase varies greatly in different parts of the world. South and Central America have the highest annual rate of increase, at present about 3 percent, followed by Africa, Asia (exclusive of the U.S.S.R.), and Oceania. More than half of the world population is Asian, 1,988 million in 1969. Next highest total is for Europe, 460 million. On the other hand, the rate of population growth in the United States and several other industrial countries is going down, but not sufficiently to cause a decrease of the population size.

These figures are generally accepted estimates. It is not possible to state exact population figures for either present or past time. A world population census does not exist, and in many countries population estimates are still no more than guesses. However, according to the United Nations Demographic Yearbook, the global population reached a total of 3.552 million by July, 1969.

The demographic explosion has been discussed so much that people become immune to its challenge. They cannot comprehend the overwhelming truth that 25 percent of all the humans who have ever lived are now alive and that the figure will be 50 percent within a few decades.

The population explosion is in reality a series of accelerating explosions. If the current rate of increase continues, man will rapidly approach a critical stage of survival in a world of destroyed and exhausted natural resources. Human overpopulation is in fact the worst and basic form of environmental pollution. It has been calculated that within 260 years the world population will be 400 billion, which corresponds to a global population density like that presently prevailing in London. Another estimate predicts that in about the year 2400 there will be about one square yard of land for each human individual on earth. But by then the population

Growth of World Population to the Year 2000

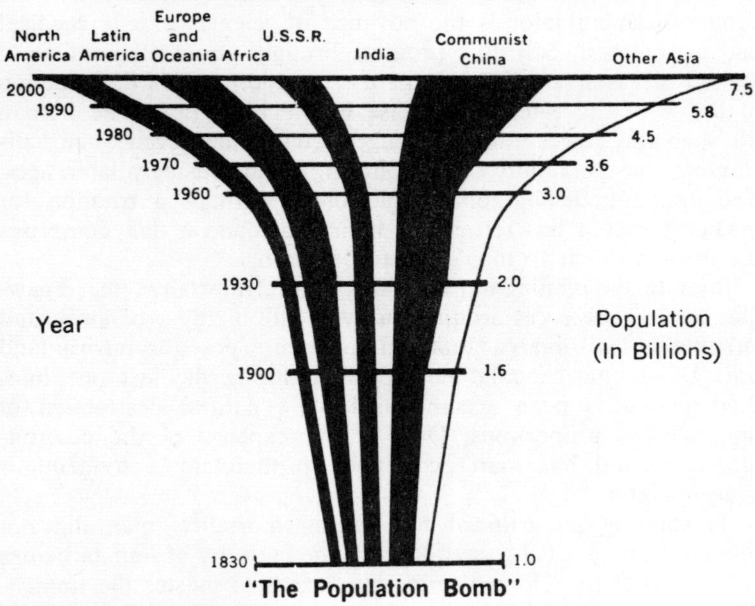

Source: Department of State.

will probably long since have exceeded the limits of man's tolerance of other members of his own species.

Birth rates and death rates are the regulating factors on populations. When they were controlled by natural means, the human population increased slowly or was rather stable. In heavily overpopulated India, for example, there was no net increase of the population prior to 1850. The natural environment checked the population size through a high mortality, which in its turn was related to the carrying capacity of the environment. Similar situations prevailed all over the world. However, such natural population control did not prevent local overpopulation in areas where the number of people exceeded their possibilities of subsistence. This was the case, for instance, in the Mediterranean countries and in the man-made deserts of all continents. Today we are facing world overpopulation with local and especially critical peaks in southern Asia and Latin America. In Asia the crisis is in full development because the disparities between the population size and available natural resources become more pronounced with each new day.

Many factors underlie the sudden population increase, but the common denominator is the advance of science, chiefly medical and agricultural. Scientific progress brought about elimination of several diseases, amelioration of the nutrition standard, reduction of the mortality rate, an increase in the birth rate. The human life span was considerably prolonged, giving man, even when half-starving, the possibility of reproducing at increasingly later ages. The dramatic demographic explosion is perhaps a triumph for medicine, but it has released a human avalanche that comprises the gravest threat to man's future on earth.

Prior to the elimination of man's natural mortality, the renewable natural resources around him were still highly productive and variable, at least in areas that had not been exposed to unwise land use. The human population explosion during the last one hundred years has been accompanied by a natural destruction of unparallelled proportions. Only at the expense of the environmental capital has man been able to maintain a tremendous overpopulation.

Substituting an artificial for natural mortality, man did not foresee the ecological consequences. The majority of human beings still fail to do so. No ecological plan exists to master the tremendous population growth man has released. Already the world population exceeds the available food supply and carrying capacity of the most densely inhabited lands. The human needs of living space have been ignored. Further population increase automatically leads to a lowering of environmental quality and living standards for most people.

The danger of failure to understand our own population problems is that every day it becomes more difficult to tackle them on an ecological basis without serious social dislocations. The fact of 129,600 births a day illustrates the dilemma. This daily rate is increasing while death rates are declining with a significant increase in human longevity. However, during recent years death rates have tended to go up in many parts of the world, an implication that man ultimately depends on the stability of ecological ecosystems, which today collapse due to man himself.

The population problem is a much debated issue. Birth control and increases in food production are often offered as the main means for solving the situation. Efficient planning of regional population sizes through birth control undoubtedly is the basic measure to be undertaken. Obviously such planning essentially must relate to the capacity of the renewable natural resources to supply continuously what the population needs without detriment to the environment. But there are many other factors to consider. The structure of a regulated population is of extreme importance for the future of a people or a nation. Some populations have to be reduced or remain static. A very few others may need to increase. The problems of dealing with regulated populations that have been deprived of natural dynamics may turn out to be extremely complicated, but they must nevertheless be dealt with.

Most difficult—for ethical, sociological, religious, and traditional reasons—is the question of how to put control of populational size into efficient practice. There are several direct methods of birth control—natural, mechanical, chemical, and abortive. A discussion of these means is beyond the scope of this book. The important fact is that birth control is possible, but it must be based on education, social means and voluntary cooperation without being forced upon people by legislative decrees that are impossible to implement. Social measures in the form of lost privileges for families having more than two children have been tried in some countries. However, the moral, religious, and traditional dimensions of the birth control problems are still enormous. Fortunately, there is an increasing tendency in both developed and developing countries to realize that there is an intrinsic relationship between environmental deterioration and population increase, which leads to increasingly favorable attitudes toward population limitation. Research is continuing on improved methods of contraception with no adverse side effects.

Biologists must activate the politicians about the necessity of birth control. An important step in that direction and an action of distinct significance is what happened on March 10, 1970, at a meeting of the Governing Board of the American Institute of Biological Sciences. Its Study Group on Public Responsibilities had drafted a resolution, which was passed overwhelmingly:

> Whereas scientific study has clearly identified as a threat to human life and to its quality the current rate of population increase and consequent overpopulation both in the United States and elsewhere, and whereas immediate measures must be taken to correct the population growth in the United States, therefore be it resolved that it is essential that the government accelerate its efforts toward implementing all methods of humane birth control at the earliest feasible time. Resolved further that the Governing

> Board of the AIBS endorses the action of Senators Packwood, Tydings, and others in Congress for their efforts to control population growth.

The bill that was proposed by former Senator Tydings (with 21 co-sponsors) was to establish a National Center for Population and Family Planning.

The bill introduced by Senator Packwood was a landmark in this field, because it suggested a departure from strictly voluntary measures and existing law by proposing to amend the Internal Revenue Code so as to restrict to three (after 1973) the number of personal exemptions allowable for children.

The Family Planning Act was passed by Congress in 1970. Voluntary birth control services would be provided for any woman who could not otherwise afford them. These services would include birth control pills and other contraceptives, but none of the funds appropriated could be used in programs where abortion was a method of family planning. In the international field the United Nations General Assembly has designated 1974 as World Demographic Year to promote the realization of birth control programs.

Socioeconomic controls are perhaps efficient tools for curbing the population growth in developed countries, but in the most critical areas of the world as far as human populations are concerned they may probably not work.

There, other systems of preventing population growth must be adopted.

The British scientist A. S. Parkes (1969) has drawn attention to the fact that the Universal Declaration of Human Rights, issued in 1967 by the United Nations, makes only one reference to reproduction, and says nothing about the consequences for the human community of unlimited reproduction by individuals. The Declaration makes many references to the political environment, but contains not a word about the problem of living space in the world or the fact that man is subject to biological laws. He should also have ecological rights. As a biologist Parkes suggests an addition to the Declaration of Human Rights in the form of a "Declaration of Human Obligations," one article of which would state:

> It is an obligation of men and women:
>
> (a) Not to produce unwanted children.
>
> (b) Not to take a substantial risk of begetting a mentally or physically defective child.
>
> (c) Not to produce children, because of irresponsibility or religious observance, merely as a by-product of sexual intercourse.

(d) To plan the number and spacing of births in the best interest of mother, child, and the rest of the family.
(e) To give the best possible mental and physical environment to the child during its most formative years and to produce children, therefore, only in the course of an affectionate and stable relationship between man and woman.
(f) However convinced the individual may be of his or her superior qualities, not for this reason to produce children in numbers which, if equalled by everyone, would be demographically catastrophic.

There can hardly be any "human rights" at all in the world if there are no "ecological rights" for man. The Preparatory Committee for the United Nations Conference on the Human Environment in 1972 will cover that point. It is preparing a draft for a "Declaration on the Human Environment."

Economy and Technology

The myth has steadily grown that man can continue indefinitely to transform his environment by technological innovations, and by doing so transform himself. This is a false and dangerous assumption. For lack of ecological knowledge, the natural resources are being planned, manipulated, and "developed" almost exclusively on the basis of technological and economic criteria, plus short-sighted political considerations with little regard for ecological, biological, and long-term social effects.

The spectacular power and feats of modern technology in the advanced countries have prevented people from seeing beyond the impressive facade of cultural and industrial revolutions. Nuclear power plants, electronic machines, synthetic chemicals, agricultural and drainage schemes, or dam-building for irrigation or hydroelectricity, which transform entire landscapes and ecosystems, have given man the impression that he masters his environment by technological means. The undeveloped countries want to achieve the same results, and that is what modern technology offers them. Ambitious programs are being planned for Africa, Asia, Latin America, and Oceania by a number of industrial countries and international organizations. They suggest agricultural or industrial development or reorganization, the transformation of river basins, or exploitation of virgin forests. The usefulness of all these projects is doubtful. Many of them have been or may be failures, bringing serious ecological consequences in their wake, because technologists, economists, and politicians continue to ignore, or avoid taking into account, the character of the ecosystem of the areas concerned. They forget the basic and the most important part of the project.

The methods of exploiting resources often create problems which are greater than the value of the resources themselves. This is due to lack of ecological awareness.

An impressive series of past failures show how environmental deterioration follows upon great development processes that interfere with the functioning of ecosystems. Some of these case histories were discussed at a conference on the "Ecological Aspects of International Development" held in Virginia in 1968. Russell E. Train, then president of the Conservation Foundation in Washington, D.C., later U. S. Undersecretary of the Interior, and at present chairman of the Council on Environmental Quality in the Executive Office of the President of the United States, made this statement:

"The adverse environmental consequences of much well-accepted technological progress are perhaps most readily and dramatically seen in international development programs where alien technology and alien goals interact with a traditional culture and values. Developing countries are defenseless before the self-assured wisdom of Western planners. We have a very heavy moral obligation to assess the full range of consequences of those international development programs, both bilateral and multilateral, which we have undertaken so confidently."

Robert Cahn has summarized from the conference discussions the harmful side effects that environmental aid programs are producing in different parts of the world. The following are examples.

In South America, fishing of the anchovy stocks off the coast of Peru, financed by outside capital, has reduced the basic resource from catches of 26 million tons to 12 or 13 million tons per year. Because of inadequate regulation, the catch in 1967 was about 11.5 million tons and in 1968, 11.2 million tons, and the resource continues to diminish.

Although even now Peruvian-Chilean fishery production exceeds the total meat production of protein-deficient South America, it is being used almost exclusively to feed livestock in wealthy countries, especially in the United States, where the fishmeal is used to fatten chickens for market.

In the Canete Valley of Peru, widespread use of insecticide was encouraged in 1949 by American development interests to eliminate pests and increase cotton yields. Seven years later, the cotton crop had gone down 50 percent, and destructive insect species had doubled. This pesticide fiasco was an aerial-spraying program using materials developed by United States industry and advocated by the U. S. Department of Agriculture, which had tested them. The insecticides caused total destruction of useful insects along with the parasites, and brought about a breakdown of the natural equilibrium within the ecosystem. The pests gradually developed immunity and increased in numbers to a point of producing tremendous damage to the crops. In 1957, the insecticide program was dropped in favor of biological control, which has been successful.

The island of Banaba in the Central Pacific was recently abandoned by its inhabitants after most of the soil had been removed, and the water situation had deteriorated, due to phosphate mining operations. The factors causing this ecological bankruptcy were as follows: On the raised limestone islands such as Banaba in the Central Pacific, mining removed most of the soil. Agriculture became impossible without elaborate procedures to convert coral rock to soil. Then the water situation, always marginal, deteriorated because of the absence of vegetation that previously had held water. Life became so difficult that in spite of the islanders' traditional love for their home island, they were forced to use the shares of phosphate profits they had saved up to buy an island elsewhere and move to it.

In Asia the effects of the Mekong River project have already greatly altered the ecosystem of an enormous area of the southeastern countries of the continent. Yet the United States government has offered help in building a 325-foot-high dam in the Mekong in Laos, which would create a reservoir twice the size of Lake Mead on the Colorado in Nevada. It would have a power plant with a capacity two and one-half times that of the Aswan project on the Nile in Egypt. It would supply irrigation water for up to five million acres of land in Thailand and Laos, plus supplemental water for another seven million acres in the Mekong Delta in South Vietnam and Cambodia.

Not nearly enough is known about many possible effects of the Mekong River development. There is clearly some one-sided emphasis in the planning. Increased power and improved agriculture are desirable, as are the other benefits of dams, such as flood control and the improvement of navigation. But disfunctions in even the nonhuman ecology may result from a planning process that relies too much on what man can do with bulldozers and cement. Strains in the relationship of man to the land on which he has to live are inevitable when such rapid changes are initiated.

Thus the construction of a dam on a major river system to produce power for industrial development may add to the serious ecological repercussions already present along the river as a result of the project's environmental impact. It would lead to increased loss of rich alluvial land behind the dam, to further displacement of people, to increased loss of organic nutrients that once naturally fertilized agricultural lands below the dam, to increased loss of fisheries that depended on upstream nutrients, to fostering the spread of disease borne by proliferating aquatic snails.

Another typical case from Asia, although not mentioned at the Virginia conference in 1968, is exemplified by the Indus Valley in West Pakistan. There the population grows at the rate of ten more mouths to be fed every five minutes. In that same five minutes in that same place, an acre of land is being lost through the water-logging and salinity.

A classical example comes from the Great Lakes region in North America, where the St. Lawrence Seaway has contributed to economic growth but also caused high cost of environmental capital. The completion of the canal between the sea and the lakes let the parasitary sea lamprey into the Great Lakes. Trout, which had been the backbone of the lakes' fishing industry, suffered greatly from the lamprey invasion. By the mid-1950's the trout and some other large, commercial predatory fish were nearly extinct. And with their near-extinction, smaller fish, especially the alewife, normally kept under control by these predators, proliferated. The aggressive alewife dominated the food supply and greatly reduced the numbers of the remaining smaller native fish, such as the lake herring. The alewife became so numerous, in fact, that on occasion great numbers died and the dead fish along the shore caused a major public nuisance.

Man attempted to restore the ecological balance by instituting sea lamprey control in the 1950's and 1960's and, beginning in 1965, by stocking the lakes with coho salmon to replace the lost native predatory fish. Feeding on the abundant alewife, the salmon multiplied rapidly and by 1969 had become important both as a commercial and sport resource. Some of the salmon, however, were contaminated by excessive concentrations of DDT and were taken off the commercial market.

The lesson is not that such activities as the St. Lawrence Seaway must be halted, but that the consequences of construction must be carefully studied in advance of construction. Planners and managers must begin to appreciate the enormous interrelated complexity of environmental systems, weigh the tradeoffs of potential environmental harm against the benefits of construction, look at alternatives, and incorporate environmental safeguards into the basic design of new developments.

When the sluice gates of the first dam across the Zambezi River in central Africa were closed in December, 1958, the rising waters began the formation of 1,700-square-mile Lake Kariba, now the largest man-made lake in the world. The first of the major African impoundments, it altered large sections of African landscape as well as the lives of people in many parts of Rhodesia and Zambia. Millions of dollars were spent on economic and engineering studies. Yet prior to the decision to proceed with construction, not a single environmental survey of the lake basin or citizen relocation areas was initiated.

A variety of unforeseen disruptions have occurred since the dam was completed. The planners had expected that the increase in the fishing resource would compensate for the loss of agricultural lands that were inundated. But after a temporary rise in the fish catch, yields dropped drastically due to factors not fully understood. The expected cash income and jobs did not materialize. Creation of the large lake shore also produced an environment highly suitable for the tsetse fly, which fostered serious outbreaks of disease among cattle. Many of the people dependent upon cattle for their livelihood have felt the economic impact. Because the dam was built primarily for hydroelectric power, and water is released for that purpose, shifts in lake level and effects on the river area below the dam have been irregular, bringing considerable uncertainty to farming efforts. In addition, the planners did not take into consideration the inadequate amount and quality of land available for resettlement of the displaced population, or the people's unwillingness to change their long-established land use practices. These practices have led to erosion and destruction of the soil in the new areas.

Another consequence of the man-made Lake Kariba was a series of earthquakes. These shocks were caused by the weight of water reactivating old fault lines along the Zambezi River and around the Kariba Dam. The seismicity of areas designated for large dam projects should be closely investigated.

An unplanned side effect in the Mediterranean of building the Aswan High Dam in Egypt and Sudan is that the North African sardine industry practically collapsed because the Nile has ceased to fertilize the marine waters outside its delta, causing a loss of about $7 million a year.

All these cases of deterioration caused by technology show,

together with many others, that development has tended to sacrifice basic environmental values to short-run economic and political expediency. The attitude toward ecological consequences, if they are even considered, is to meet them if and when they arise. But at that stage ecological damage has already occurred, and it is often too late to call upon ecologists for advice. The solution must be increased attention to ecological factors as an integral part of all phases of the development procedure.

In this technological age it is more and more necessary for man to understand the living world in which he lives. Depletion of natural resources should be of particular concern to a technological society which basically depends upon such resources for energy and for materials. The dramatic reality is that tremendous environmental changes wrought by technology proceed throughout the world without serious attempts by society to control what is happening behind that glittering facade called progress. Most people are not aware of the danger that lies in dissolving ancient relationships with nature through an excess of technological dynamism. They do not seem to be conscious of the increasingly rapid changes taking place in the natural environments around us. The magnitude and rate of that change within our own generation must be compared with tectonic and volcanic forces, or climatic effects much like the ice ages. Other comparisons have been made by scientists of various disciplines. An archeologist claims that during the twentieth century we are experiencing changes equal to those of the preceding five thousand years. A Nobel Prize laureate asserts that the closest parallel to today's changes in the entire history of mankind lies in the invention of agriculture in Neolithic time. An atomic scientist calculates that half of all the energy consumed by man in the past two thousand years has been used during the past one hundred years.

For some twelve thousand years all the major achievements of civilization have been closely linked to man's relationships with the biological environment. Until the present generation, four-fifths of the human world population lived in villages that were closer in every respect to the Neolithic community than to any contemporary metropolitan complex. The Neolithic culture, it is true, never reached the heights of civilization that urban societies did during the Bronze, Iron, or Nuclear Age—but on the other hand it never descended to the depths of destruction that characterize modern times.

In the industrial countries rural-based life decreases rapidly and the richest soils and the most important cultivable areas are transformed into urban deserts by the growing cities, expressways, parking lots, and airports at the same time that the air, the water, and the soil are increasingly polluted. Every year the urbanoid mass of United States' cities expand into more than a million and a half acres of open space, much of it excellent agricultural land. In California alone it has been estimated that within thirty years at least half of the best cropland will have been transformed to urban and industrial use. All these changes caused by the technological revolution lead to biological danger for man. Ultimately it will become a question of survival.

Modern man's tremendous consumption of energy is another danger sign of ecological unbalance due to technology. The latter tries to overcome organic limitations without being equipped with the self-regulating devices that organic systems have developed. This is another example of man's attempts to go beyond his biological limits.

It is only within the last decades that man has used energy in concentrated form, but in about three or four hundred years he will have consumed most of the world's fossil fuels that were produced and stored by the earth over the last several hundred million years. A global energy crisis is already visible at the horizon. Nobody seems to plan for it, perhaps because technologists and politicians hope that nuclear power will solve the problems.

The drive behind technological progress is economic profit, and the notion that an increase in the standard of living can be measured only in economic terms. People too often forget to ask: whose progress? The shortcoming of a philosophy based on economic expansion is its neglect of long-term environmental processes and failure to understand the risks involved in applying technological machinery to reshaping the earth. It has produced an overheated technology that each year drives faster at the expense of environmental health, due to lack of ecological insight and foresight.

Politicians, economists, and technologists seem dogmatically to believe—or at least they try to convince people to believe—that the size of the Gross National Product (GNP) determines the quality of life independent of how far the environmental destruction goes on. It is high time to adopt newer criteria for measuring national well-being than the conventional indices of growth of the GNP which are often based on a consumption and disfunction of renewable natural resources, or in other words, capital.

The present economic system, capitalist and communist, must be exposed to a rigourous examination based on ecological and sociological considerations before it is too late to avoid disaster. We must act now. Our generation may be the last to be offered a free choice concerning the future of mankind, and the last able to undertake a long-term policy to restore the natural environment with any prospects of success. One thing is absolutely clear: there is no technological solution.

Fortunately, public opinion has begun to react, and some governments are starting to understand the true significance of the issue. In 1969 the Gallup organization took a nationwide poll to determine the attitudes of the United States' public toward their environment. The replies to two questions are of special interest. One question was, "Which of these kinds of places would you find most pleasant as a place to live?" Judging by the replies, it is clear that if a pleasant place to live were the principal consideration influencing the public, there would be marked reversal of the trend away from rural areas to technological megalopolises. The results divided as follows among those interviewed:

Place	*Percent*
Rural areas	30
Small city	25
Suburbs	18
Mountains	15
Seashore	9
Large city	6
Other	1

The other question was, "In this country, which one of these do you think is the most pressing problem connected with our natural surroundings?" Those interviewed ranked the relative urgency of environmental problems as follows:

Problem	*Percent*
Air Pollution	36
Water Pollution	32
Pesticides (chemicals used to kill insects)	7
Preservation of open green spaces	6
Wildlife preservation (birds and animals)	5
Soil erosion	4

Environment and Development

The developing countries are eager to reach as soon as possible the same degree of development and standard of living as the industrial countries. Can such a goal be achieved without environmental disruption? It certainly can, if the political will is firm to avoid unnecessary damage to renewable natural resources by maintaining environmental quality as an integral part of the development process. Such a development program will be costly at the initial stage, but will pay off in the long run.

However, the environmental threat to the developing countries comes actually from their present misuse of water, soil, vegetation, and wild animals. These countries are in reality using up their renewable natural resources before the development process has accelerated or even begun. This dramatic situation is caused by the population increase and lack of control as to how the land is used. It is a terrific challenge for a nation, because it undermines its potential for development. None of the developed countries have been in the same dilemma, because there the development process was based on intact natural resources.

Radioactivity

Nuclear power is another technological triumph having an immense value for mankind but bringing with it tremendous environmental risks, both short-term and long-term. The effects of radioactive fallout are global. The problems of how to dispose safely of enormous amounts of radioactive wastes have become increasingly difficult. Many countries are dumping dangerously toxic by-products of atomic energy plants into the sea. In the United States radioactive wastes have been stored in concrete blocks buried in the oceans, caves, salt domes, and deep wells. There is no guarantee that containers dumped in the ocean or placed in terrestrial deposits will remain intact forever. Even greater is the risk of accidental release of radioactive material from ships using nuclear power, or from aircraft carrying atomic bombs. The effects of radioactive contamination in a freshwater system or in the oceans would constitute an unprecedented danger to all life forms.

Fortunately, the period of experimental nuclear explosions in the air seems to be over. The consequence of the Russian atomic explosions became very quickly apparent in the Arctic because in the Far North food chains are short, and the ecosystems relatively simple. Radioisotopes of strontium, iodine, and caesium, which are persistent up to twenty-eight to thirty years, were carried by air currents to the Arctic and deposited there. Lichens absorbed the radioactive fallout, and caribou and reindeer that fed on the lichen acquired a heavier load of radioactive fallout than any other species in the world. Eskimos and Lapps, as well as others, eat caribou and reindeer meat, so the fatal consequences to man are direct, which is shown by the load of strontium in Eskimos and Lapps.

Despite increasing nuclear power stations in many countries, the medical applications are still the largest source (94 percent) of man-made radiation to which the average person is exposed. There is also a natural radiation to which man is normally exposed. It is evident that the storage and disposal of radioactive waste will become an increasingly serious problem in many parts of the world, where the use of nuclear power will expand.

Genetics

The technological revolution and its environmental implications may have genetic effects on man, and these may not be apparent until ensuing generations. Persistent toxic chemicals that poison the environment through food chains may bring about evolution of less productive mutants in various organisms, including man. Genetic effects of radioactive fallout are another danger.

The greatest threat to the quality of the human species is the genetic effect of the unnatural expansion in numbers. Man artificially keeps individuals alive who in natural populations would have been eliminated by environmental factors. In nature, weak individuals or inefficient members of a community do not survive. In this way the species evolves naturally and maintains its quality. In human societies of today there is no biological control on defective individuals, who reproduce as frequently as healthy ones and often transmit their organic disabilities to their offspring. Again an environmental barrier and a biological process have been artificially removed, and the effect is definitely not to the advantage of the species. Evolution through natural selection has been eliminated. We face here a delicate biological and ethical problem. No species other than man is confronted with this dilemma.

Recreation

At the World Wildlife Fund's first international congress, called "Nature and Man," held in Amsterdam in 1967, a declaration was issued. It began with the following words:

> That the natural world with its infinite variety of landscape, animals and plants, and its infinite capacity to uplift the human spirit, is essential to the well being of all people and is a part of the heritage of all people, deserving the respect of all people . . .
>
> That for these reasons all people of each generation all over the world have a responsibility of trusteeship to defend and nurture these irreplaceable natural resources and hand them on undamaged to the next generation . . .

The reference to nature's "infinite capacity to uplift the human spirit" emphasizes how important recreation in a living landscape is for man, providing he has leisure time at his disposal and the means to utilize it.

The recreational value of nature is an established fact on all continents. Fortunately, it goes together with other natural values. It is not a luxury that only a few can afford. These values were listed in the introduction to the Ninth Technical Meeting of the International Union for Conservation of Nature and Natural Resources (IUCN), which dealt with the Ecology of Man in the Tropical Environment, held in Nairobi in 1963, as follows:

> Ethical–taking note of man's responsibility to preserve nature in its manifold variety.
> Aesthetic–for inspiration and repose.
> Scientific–to know nature and her dynamic processes.
> Educational–to understand the environment in which we live.
> Recreational–for change and enjoyment.
> Economic–for the material benefits derived from natural resources.

Ethical and aesthetic values form the background for recreational values. Hence the importance accorded to recreation as an argument for sound conservation of nature is well recognized not only by the international conservation organizations but also by a number of governments on all continents. However, problems often arise relative to outdoor recreation, because its impact on nature can be so great as to damage the natural resources of a particular area. The great flow of visitors to a nature reserve can destroy the attractions for which the area was preserved. Some governments want, for social reasons, to provide outdoor recreation with various facilities for the masses, without hesitating to sacrifice fragile habitats that could have remained attractive if managed properly and visited with care.

The issue of tourism and recreation in natural environments is not simple and requires careful planning. Areas that are overcrowded on holidays and are placed under such heavy pressure by visitors that they change character and deteriorate are no longer natural, and therefore lose their recreational value. They are transformed to public playgrounds and come to have little or nothing in common with nature reserves. Such a development is destruction of nature.

To a growing number of people outdoor recreation means visiting natural areas with a diversity of habitats, where one can follow the seasonal events in the plant and animal worlds. "Wildlife watching" was the most popular single recreation among an estimated 15.6 million visitors to the U. S. national wildlife refuges in 1967. It accounted for 6,100,000 visits to the refuge system—317 units comprising a total of 29 million acres. Other people want to fish and hunt. In the United States more than 30 million people spend billions of dollars each year on fishing and hunting. They need space for their recreational activities. Landowners often find that it is more economical to convert their farms or ranches to hunting grounds or wildlife watching areas. Such outdoor recreation areas must exist both close to cities for weekend use and more remote and in varied types of biomes for longer vacations. For recreation alone, but also for a number of other reasons, governments must pursue the objective of quality environment as a top national priority that takes precedence over other, often competing, uses of environment.

National Parks and Nature Reserves

In the 1969 Gallup Poll in the United States, three of every four people interviewed favoured setting aside more public land for conservation purposes such as national parks, wildlife refuges, bird sanctuaries, and similar nature reserves. It is understandable why this pronounced enthusiasm for more national parks exists in a country like the United States that pioneered conservation by setting aside nature reserves and today has an impressive network of national parks, state parks, and other reserves visited annually by millions. Canada has followed the example of its southern neighbour. Hence North America is relatively well equipped with national parks and nature reserves.

In Africa today there are more national parks and nature reserves than during the colonial time. The new African Convention for Conservation of Nature and Natural Resources, signed in 1968 by the African heads of state and now in force, deals specifically with national parks, nature reserves, and other conservation areas. An article of the Convention stipulates that the contracting states shall maintain and extend, where appropriate, within their territory and, where applicable, in their territorial waters, the conservation areas existing at the time of the entry into force of the present convention, and, preferably within the framework of land use planning programs, shall assess the necessity of establishing additional conservation areas in order to:

1. protect those ecosystems that are most representative of, and particularly those that are in any respect peculiar to, their territories;
2. ensure conservation of all species and more particularly of those listed or to be listed in the Annex of this Convention. (At present more than 350 species of animals and plants are listed.)

The number of national parks in Europe, Asia, and Australia is also increasing. However, many more such reserves are needed, for many reasons. Two reasons that are particularly important, and therefore emphasized by the African Convention are: protection of representative ecosystems, and protection of endangered species of plants and animals. This holds true for the whole world and includes terrestrial as well as marine biomes.

National parks and nature reserves provide a number of advantages to human society, in addition to recreational values already discussed. Tourism is a kind of recreation, but as an international phenomenon it brings foreign money into countries. It can therefore be developed into a major source of income based on the existence of national parks. This is the case in the countries of East Africa. In Kenya tourism is the most important source of national revenue, competing with the value of the coffee export, and ranks high in that regard in Uganda and Tanzania. The stream of tourists visiting these countries comes primarily to see the national parks and nature reserves. Primarily it is the animals and the scenic landscape features that attract visitors and have the strongest appeal for foreign tourists. Hence the recreational and economic values of national parks and their wildlife can contribute substantially to the economy of a country.

The economic potential of national parks and nature reserves is often much greater than what the same areas can yield if exploited for agriculture, forestry, or industrial activities. A comment in the Current Affairs Bulletin on "Tourism—Today and Tomorrow" published by the U. S. Department of Commerce is pertinent here: "If the community can attract a couple of dozen tourists a day throughout the year, it would be comparable economically to acquiring a new manufacturing industry with an annual payroll of $100,000."

The physical and mental health of people is an important economic factor in any country, even though it cannot be expressed in figures. The recreational value of natural oases like national parks and nature reserves in a world of artificiality, urbanization, and mechanization, for maintaining the physical and mental welfare of human beings cannot be overestimated.

Ethical and aesthetic values are also represented by national parks. For many people it is a great satisfaction that untouched and other natural areas, and endangered species of animals, exist somewhere in a changing world. In addition, the historical aspect should be stressed. Even as national parks represent dynamic landscape entities in evolution, they are also parts of a country's past, showing succeeding generations how the land looked in their forefathers' time. It is difficult to understand why governments and societies for the preservation and management of natural sites spend only a fraction, or nothing at all, of the tremendous sums paid out annually to maintain monuments and historic buildings. This discrepancy must be due to overestimation of self by man: What he has created is worth so much more than what nature has created through evolution in the course of eons. Yet monuments and buildings can be reconstructed, while exterminated plants and animals cannot.

The educational value of national parks is high, not only because youngsters—in fact, all age classes—can learn what living species of plants and animals are like in the wild but also what ecology is. National parks and nature reserves show more clearly than any museum or textbook how a living landscape functions: the complex interrelationships between air, water, soil, plant and animal species, the function of predation, the diversity of natural lands in ecological balance, and so forth.

To this must be added enormous scientific values. Modern science needs large undisturbed areas for long-term research. Sample areas of all habitats and ecosystems with intact biocommunities are necessary for comparing productivity and evolution with cultivated areas or man-made habitats. Here again it is of direct economic interest to any society to set aside a network of intact natural areas for scientific research. The dividends from such measures cannot be foretold, but may lead to discoveries of great importance for the future of mankind.

The manifold economic values of national parks and nature reserves have been emphasized many times in this book. But even if such areas and their wildlife did not have those values, they would still be worth preserving for their human inspirational value alone. Even though important functions of national parks are recreational, educational, and scientific, the prime purpose of such reserves is to preserve. Conservation and management of protected, natural areas should be the first goal of any true national park. Often recreational uses and overutilization by too many visitors come into conflict with these goals and even damage the values for which the reserve was originally established. On the other hand, national parks must remain popular in the sense that people like them, use them, and want more of them. Therefore a compromise should be sought. It is a matter of management. In general, it may be justifiable to channel visitors by roads and other arrangements to easily accessible and representative areas of a national park, providing this does not damage vital parts of the reserve and that about 90 percent of the reserve remains intact, without roads and facilities other than minor trails and modest huts for hikers. This system of zoning has been shown to function satisfactorily in many of the world's national parks. Many animals soon adapt to tourist conditions, while the shyer and more sensitive ones stick to the wilderness zones.

Curiously enough, the fish fauna and aquatic habitats in national parks are often treated in another way than terrestrial animals and habitat. Principally this is difficult to understand. Stocking of exotic or native fish within a national park is an artificial manipulation which may destroy the integrity of aquatic ecosystems and therefore ruin an important part of the reserve.

The views on national parks and nature reserves expressed in this section are today shared by United Nations organizations. At the Intergovernmental Conference of Experts on the Scientific Basis for Rational Use and Conservation of the Resources of the Biosphere, organized in Paris in 1968 by UNESCO, the discussion on national parks and nature reserves emphasized that such preserved areas are of great economic and scientific importance. The creation of national parks and nature reserves often implies a choice between conflicting national interests, and this choice should be preceded by widespread consultation held only at a national level. It was agreed that wherever no national organization yet exists to select, establish, and manage a network of national parks and nature reserves as part of an integrated plan, such an organization should be established at an early date. This organization, whether already in existence or yet to be created, should work out or further elaborate a national program for national parks and nature reserves, endeavouring as far as possible to conform to the selection criteria and standard nomenclature set up in accordance with resolution 713 (XXVII) of ECOSOC, and in accordance with the principles laid down by the Secretary-General of the United Nations. These are the criteria and standards that IUCN recently used for the publication of the second edition of "United Nations List of National Parks and Equivalent Reserves."

It is obvious that the integrity of national parks must be respected, particularly by the governments that have established them. Nevertheless, too often this is not the case. The most outstanding examples of unnecessary damage to national parks and ecosystems come from Sweden, where several national parks and nature reserves have been violated by wholesale governmental exploitation during the last twenty years—despite the fact that this country is one of the richest in the world, one of the largest in Europe, and has a very small population with an extremely high standard of living. Lack of vision, of conservation conscience, of ecological responsibility, of sense of duty toward future generations, combined with blindness to greed and political coveting of wealth and power, seem to cause some governments to behave in the tradition of *"après nous le déluge."*

Research

Human societies have become so large and use renewable natural resources so rapidly and in such a complex way that they are in danger of being drowned by their own pollution. Only intensified ecosystem-oriented research on a global scale, involving all components of the environment, may create solutions to innumerable problems vital for human survival.

The International Biological Program (IBP), originally an idea of the IUCN, developed by the International Union of Biological Sciences (IUBS) and the International Council of Scientific Unions (ICSU) and now an entire organization in itself, is working along these lines. The aim of the IBP is research devoted to biological productivity and human welfare. It is a parallel to the International Geophysical Year, and has been carried out during a six-year period (1967–72). It will be taken over and followed by a long-term program of even greater magnitude, "Man and the Biosphere" (MAB), initiated and organized by UNESCO in collaboration with other UN agencies and IUCN.

These programs are very ambitious, but they also are realistic, covering a global spectrum of problems that are biologically significant—for example, chemical cycles in living systems, transport of airborne spores and pathogens, human responses to stressful environments, pollution as an obstacle to productive systems, colonization of islands, human genetics and nutrition, reactions of disturbed biocommunities, and so on. All these topics are, in fact, parts of an analysis of ecosystems, which is precisely the central topic chosen by the United States IBP Committee. It is a wise choice, because a better understanding of the function of ecosystems is badly needed. The problem is vast and complex, but central indeed to man, who is an integral part of most ecosystems, not only influencing them but also being influenced by them.

Ecosystem research was discussed at length at UNESCO's 1968 Intergovernmental Conference of Experts on the Scientific Basis for Rational Use and Conservation of the Resources of the Biosphere. The conferees particularly stressed the importance of research into the functional aspects of ecosystems. Their discussions underscored the fact that knowledge of how ecosystems function is scanty, in spite of the fact that it forms the essential conditions for good management. Such research, it was agreed, should be based on both the global and the analytical approach. The global approach would comprise studies on total biomass and its variations with time, energy flow, efficiency of energy conversion (photosynthesis, energy conversion between trophic levels), evaluation of input and output of energy and mass, as well as international transfer processes. These studies would emphasize nitrogen fixations, biogeochemical cycles, soil biology, biogeochemical activity of living organisms, and water-soil-plant interrelationships. The analytical approach would take into account problems of response of single individual plants and other organisms to changes in external and internal factors, and would seek to establish the laws governing the exchange of energy and mass between organisms and the environment. Conferees also stressed that research is needed into the influence of climatic conditions and their variables on ecosystems and, particularly, on land and water use potentials.

The problem of response of the whole ecosystem to external influences—i.e., management—was brought out. Comparative studies of actual and potential production, as well as the effects of changes of structure on the functioning of the ecosystem, were recommended. Changes in the functioning with respect to time—seasonal changes, phenology, evolution of the structure and of the functioning of the ecosystem—would also be studied. A more theoretical aspect was commented on: the reduction of entropy during the evolution of the ecosystem.

In order to implement such research, it was stipulated, new techniques should be developed, and a network of research areas set up. Although IBP at the present time does useful research work on primary and secondary productivity and energy flow in ecosystems, this could only be considered as a first step toward full study of the problems involved. IBP should therefore be followed up by an intergovernmental program of biosphere research emphasizing the functioning of ecosystems in the broadest sense. Only research work on that level can give a better understanding of the consequences of interference caused by land development. Seminatural and agricultural ecosystems should be included in this scheme. Human ecology should be considered as a section of general ecology. There was general agreement that inventory work as well as ecosystems analysis is hampered by a lack of taxonomists.

Conservation Education

In almost all countries conservation is conspicuously absent from educational programs in schools, colleges, and universities. There is no educational antidote to the fact that man all over the world continues the accelerated abuse of his environment. Conservation conflicts often arise from many and diverse sources—individuals, companies, cities, counties, government agencies at all levels. On the basis of traditional antecedents, selfish interests, political ambitions, or ecological ignorance, these persons and agencies propose developments that are of doubtful value at best, and frequently are potentially dangerous in their environmental consequences.

That conservation education be included in school and university curricula and incorporated in various disciplines in order to develop ecological literacy is of utmost importance. A broad approach is necessary—multidisciplinary, but not lacking in any single discipline, particularly not the technological ones. It is surprising that all kinds of technologists, many of whom are supposed to handle environmental problems, graduate from their institutions without knowing any ecology.

Dr. Herman C. Kranzer, professor of Science Education at Temple University, Philadelphia, has taken up the educational problem in an article in BioScience (1968): "Finding livable solutions to conflicting conservation issues will put our society to one of its most severe tests. We shall have to draw heavily on the enlightened leadership of the brave new college generation to see us through the battles of environmental pollution, food and population, and other critical problems. A conservation-wise citizenry will help assure intelligent solutions through peaceful, democratic processes. Let us rise to the challenge."

The younger generations in many countries are beginning to be conscious of the conservation message because they have discovered the penalties of reckless exploitation of natural resources. Therefore many students are dissatisfied with what universities teach, or rather with what universities do not teach.

In this field of education, as in the much larger field of conservation in general, it is up to the biologists to act. Those who are academic teachers are well situated to take initiatives to break the traditional bounds of several disciplines by introducing conservation and ecology subject material. In a wider context it is likewise up to the biologists within each society to advocate ecological principles. The years ahead will certainly be a period dominated by biological thinking and ecological concepts. Therefore the present generation of ecologists have a tremendous responsibility. They must make a breakthrough in time to give society the tools and the understanding with which man can control himself, before he falls under the control of the environment.

The United States offers an example of how successfully the environment can be managed when the work is directed by biological scientists. That country has a tradition of making wildlife management largely a responsibility of biologists. The past forty years have seen unparallelled progress in fish and wildlife management and conservation, giving the United States global leadership in this field and placing it several decades ahead of other advanced countries. Much that has been achieved is due to the circumstance that all environmental problems concerning fish and wildlife have been dealt with by professional biologists, trained at the universities in such biological concepts as reproductive potential, carrying capacity, territorial behaviour, population dynamics, and other factors. Furthermore, they are also trained to apply other natural sciences such as physics, chemistry, mathematics and statistics that are important for understanding and dealing with the environment.

I do not mean, however, that biologists or ecologists alone should do this job, with its world-governing implications. While ecologists are the most competent to measure, understand, and evaluate man's impact on the environment, and to suggest means for dealing with it, human population problems must be worked out in cooperation with sociologists and economists. Ecologists need to expand their field of knowledge to include human behaviour, while social and economic scientists should recognize that man is a biological product and a part of the ecosystem.

The Ethical Aspect of Nature

Man believes that he is the only ethical animal. This may be true, but those who have watched the behaviour of such wild animals as higher primates, cetaceans, pinnipeds, and carnivores frequently see social features that suggest a code of ethics, although the behaviour is biologically motivated and has evolved through natural selection. Actually man's ethics are highly irrational. They vary not only in different societies but also among individuals of the same ethnic groups. Moreover, man's ethical concepts of his own population crisis is a biological disadvantage.

Man has an ethical imperative to protect species of plants and animals that share the planet earth with him and have lived there much longer than he has. They have the right of continued existence. They should not be exterminated—rather should be protected for their own sake. This philosophy does not prevent the notion that the surplus of plant and animal populations should be utilized by man as useful natural resources.

A growing number of people on all continents accept this ethic, despite the fact that during the last three hundred years man as a species has behaved in a way indicating he hardly has any ethical concepts. The ethical aspect of nature and its animals does not yet carry much weight in a world where everything must have an economic value to be appreciated. In the future recognition will certainly be much more widespread than at present. Raymond P. Dassmann (1964) even goes so far as to say there is good reason to believe that if we do not accept such an ethic, the future of man on this planet is likely to be short and violent. Dassmann refers to Aldo Leopold's words (1949) of a "land ethic" that implies respect for our fellow members (plants and animals) of the land community and also respect for this biocommunity as such.

CHAPTER NINE

Is Conservation a Losing Battle?

For those who for decades have fought for conservation ideas, the apathy of public opinion and the lack of leadership of governments have been discouraging. Too many biologists in every country have until recent times either neglected environmental problems or been unwilling to tackle them because they did not want to criticize governments' inaction. Many biologists, particularly in Sweden, did not become concerned until they found it opportunistic to be conservation-minded. Now they are as newcomers on the scene, riding the crest of the conservation wave. As recently as ten years ago colleagues of mine thought I was wasting time trying to fight for conservation of water, soil, vegetation, and animals. Perhaps they were right. It may be too late today to warn of our predicament, but it would be a betrayal of our world to give up the conservation battle.

Now that public opinion has been aroused, the issue is merely to get people and governments to understand what ecology means and that man is an endangered species also, that humanity must change its provincial attitude to the environment, to its renewable resources, and to the population problem.

Too often conservationists are accused of being more interested in saving animals than men. The question is not one of the beasts or us. Man cannot exist without wild animals, but the latter can easily get along without us.

There is more than ever a great role to be played by conservationists all over the world, because it is men of global, environmental understanding who have to show leadership on this globe of ours.

International Conservation Organizations

Conservation is necessarily international. Ecosystems are not limited by political boundaries. Renewable natural resources are not controlled by national legislations. Pollution of air, water, and soil cannot be tackled successfully without international research and cooperative conservation measures. Migratory animals cannot be efficiently managed if every country regards them as a national property. Global planning of conservation, management, and utilization of all renewable natural resources are necessities for human survival. These views have often been expressed by conservation organizations and by ecologists, but until recently few governments have understood that these recommendations are realistic.

In the history of conservation it was characteristically the private sector of society that advanced conservational principles long before governments became aware of the problem. This was likewise true at the international level. Even before World War I some devoted conservationists realized that it was necessary to work internationally, but the war blocked their efforts. In 1922 the International Council for Bird Preservation (ICBP) was established as the first truly international conservation organization. Ornithologists have been the pioneers for conservation almost everywhere.

In 1928 the International Office for Protection of Nature was founded in Europe, with headquarters in Brussels. It was essentially a documentation center.

Since its creation in 1946, UNESCO has been interested in conservation. In 1948 it took the initiative to create the International Union for Conservation of Nature and Natural Resources (IUCN), which since then has been the major conservation organization of the world. The IUCN is a nongovernmental agency, although governments and governmental agencies are members along with a variety of international or national private organizations and institutions. Founding governments of IUCN were Belgium, Luxembourg, and the Netherlands together with seventy-four national conservation organizations. Today the organization comprises thirty governments and about three hundred organizations representing more than eighty nations. It enjoys the support of and consultative status with the United Nations through UNESCO, FAO, and ECOSOC. Its executive board and its six permanent commissions, on threatened species, education, ecology, national parks, legislation and policy, and landscape planning, are composed of specialists representing the elite of conservationists, ecologists, and other scientists, as well as administrators.

IUCN's main purpose is to promote or support action that will ensure the perpetuation of wild nature and natural resources on a world-wide basis, both for their intrinsic cultural or scientific values and for mankind's long-term economic and social welfare. The IUCN considers that conservation is best defined as the rational use of earth's resources to achieve the highest quality of living for mankind. To further these objectives the organization promotes:

Awareness through education, so that as many people as possible may understand the value and importance of renewable natural resources, and appreciate the need to use them wisely. Activities include general assemblies, regional meetings, symposia, and the dissemination of information through communications media and through the union's own publications.

Research to discover the best measures for conservation and to advance the study of ecology upon which all practical conservation depends.

Assistance in providing advice based on ecological considerations, scientific and technical data, source material and references, and all the manifold contacts required for the coordination and conduct of practical conservation programs.

Action on a national and international scale, by enlisting the cooperation of governments and international agencies in support of conservation programs, as well as in strengthening legislation and improving its enforcement.

The IUCN main field of activity lies in conserving the plants and animals, the soil, water, air and other natural wealth that constitute the earth's basic assets; dealing with threats to the quality of the natural environment, especially of wild lands and living resources, and proposing methods by which such problems may best be resolved; and promoting action and educational measures to advance the quality of the human environment.

The IUCN takes a leading part in the protection of rare species, particularly those threatened with extinction; in the perpetuation of natural habitats for wild animals; and in encouraging the establishment of national parks, reserves, and sanctuaries for aesthetic, scientific, and recreational purposes. The IUCN is concerned at the encroachment of unregulated development upon the natural scene, and regards as of particular importance the need for ecological principles to be applied to all land use planning.

IUCN sponsors special research projects and surveys ranging in scope from studies by an individual scientist, undertaken at the request of a government, to continent-wide projects.

Because of its unique independent status as a nongovernmental body supported by member governments, the IUCN is able to operate directly and quickly in dealing with matters of international significance.

Since its foundation the IUCN has dealt directly with governments on a great many occasions, in order to draw attention to the dangers threatening the natural resources of their countries. These intercessions have invariably been received with understanding and usually followed by constructive action.

The IUCN has a special relationship with the World Wildlife Fund (WWF), providing technical evaluation and assessment of the requests for financial assistance received by the WWF from all over the world. The WWF contributes greatly to IUCN's objectives by giving substantial financial support. In fact, the WWF was formed primarily to assist the IUCN, which is the highest priority of the Fund. The initiative to establish the WWF was taken in 1961 by sixteen international conservationists, of which many were members of IUCN's executive board. In the so-called Morges Manifesto they launched a global crusade against unnecessary destruction of the world's renewable natural resources, primarily wildlife and its habitats. Since then the WWF has shared headquarters with IUCN.

The aim of the WWF is to support the conservation of nature by raising funds through national appeals and allocating them to various kinds of projects. The WWF began its operations in 1962. Up to January 1, 1972, it had succeeded in raising over $6 million, mainly through the efforts of the Fund's national appeals in the United States and Great Britain. These funds have been channelled to conservation projects all over the world.

Other priorities of the WWF are the ICBP (mentioned above) and the International Wildfowl Research Bureau (IWRB). These two organizations also collaborate closely with IUCN. The ICBP is organized in continental sections, which are divided into national sections. The IWRB is a daughter organization of ICBP, its objective being to coordinate and organize research and conservation activities concerning the populations of swans, geese, ducks, waders, coots, and flamingos. Most of these birds are migratory, and many species are economically important. Therefore it is essential that their habitats used for breeding, moulting, feeding, resting, and wintering be conserved, managed, and utilized properly. This can hardly be done without organized international cooperation.

Despite very modest economic resources, the IUCN, the WWF, the ICBP, and the IWRB have worked intensely on a global scale to save the world's heritage, to introduce ecological principles, and to awaken governments. They have provided the technical background for international conservation work at regional, continental, and global levels. The IUCN in particular has played a leading role by introducing the conservation philosophy that today forms the basis for the affirmative side of the general conservation debate. The message of this conservation thinking seems finally to have gotten across to at least some governments. But the main burden of conservation work continues to lie on the private organizations.

The fact that these international, nongovernmental conservation organizations have been able, despite great financial difficulties, to achieve some positive results in their fight to conserve natural environments is due entirely to devoted conservationists who have given their utmost for decades. Their only compensation has been a degree of conservation progress. But their struggle must go on, because more natural resources are still being destroyed than are being saved or restored.

Within the United Nations it is UNESCO that since its creation in 1946 has worked for conservation either directly or by supporting the IUCN. UNESCO has undertaken a number of conservation projects and initiatives, primarily educational. At a much later stage the Food and Agriculture Organization (FAO) took up conservation of nature. It was induced to do so after having executed part of the so-called African Special Project, which was initiated and organized by IUCN in 1960–63 and partly financed by the UN Special Fund. After completion of this project FAO established a "Wildlife and Forest Recreation Section" within the Forestry and Forest Industries Division. At present FAO runs many projects in Africa, Asia, and Latin America that are important for the management, recreational aspects, and other utilization of natural resources. Most of these projects are financed by the United Nations Development Program, which is gradually, like the World Bank, coming around to the idea that ecological surveys should precede development plans.

Several of IUCN's initiatives have led to separate international organizations in addition to the WWF and the IBP. A third is the Charles Darwin Foundation for Galapagos Isles, which is working for research and conservation in the Galapagos in cooperation with the government of Ecuador and UNESCO.

In recent years the International Council of Hunting (CIC) has become increasingly conservation-minded, cooperating particularly with the ICBP and the IWRB. Regional political organizations like the Council of Europe, the Organization of African Unity (OAU), and the Organization of American States (OAS) understand the importance of conservation. All of them have achieved results in cooperation with their organizations, particularly IUCN. Three national organizations dealing with world-wide conservation work should be mentioned here. They are the Conservation Foundation in Washington, D.C., the Fauna Preservation Society in London, and the Zoologische Gesellschaft in Frankfurt. Finally, two governments have more than any others contributed materially and spiritually to international conservation: the United States through its Department of the Interior, and Great Britain through its Nature Conservancy.

This review of international organizations dealing with conservation may give the reader an impression that there are too many bodies involved, leading to unnecessary duplication of efforts and loss of money. It is not so. All these organizations cooperate in a synchronized manner. The field of conservation of nature is so wide and variable, and there is so much to do that each of these organizations has its particular role—its, so to speak, own niche.

CHAPTER TEN

Continental Problems of Today

Although many conservation problems are universal, each continent has peculiar problems. All over the world the welfare of man is bound to the natural resources available to him. The problems have arisen as a consequence of how man has utilized and still uses these resources. The ways of land and water use differ on various continents, despite the fact that everywhere on this planet man treats the renewable natural resources as if they were inexhaustible.

Man has placed a stress upon the biosphere, and thus upon himself. Environmental management through conservation and wise utilization of natural resources are the only solutions to these man-made problems.

North America

In 1969 the President of the United States created an Environmental Quality Council, composed of six Cabinet heads that deal with natural resources. By this important step conservation attained recognition as a top issue in the life of a nation. This reflects the remarkable progress of conservation in the United States during the 1960's, when the general attitude shifted from indifference or ignorance to keen interest and the passage of significant legislation governing use of natural resources. Yet for half a century the United States has been several decades ahead of all other nations in the field of conservation. However, that is not to say there are no serious conservation problems in the United States. Too many still exploit natural resources unwisely, yielding short-term profit to themselves at the cost of long-term benefit to society.

This attitude is an environmental danger, but it may be reversed by the marked, growing public interest in conservation in the United States. It is high time for such a reversal of attitude. The pollution of air, water, and soil has tremendous proportions. According to the University of California's Scripps Institution of Oceanography, the United States appears to be responsible for about one-third to one-half of many of the contaminants introduced into the atmosphere or oceans. Levels of pesticides such as DDT in deep-living marine fish are similar to those of terrestrial organisms. Radioactivity from the detonation of nuclear devices and emissions from nuclear reactors are found at all levels in all oceans.

In acreage the United States is not even half of North America, but nevertheless the pollution problems of this country have an impact on the entire continent up to the Arctic tundras of Alaska and Canada. As in all industrial countries, environmental pollution is actually North America's number one conservation problem. The oil boom in Alaska is another important conservation matter. It may upset the whole ecosystem and break down both terrestrial and marine resources, causing far-reaching detrimental consequences if the exploitation is not carried out with great care.

Population growth will certainly become a problem in several parts of the United States, as it already is in Mexico. Due to increase and uneven distribution of the human population in the latter country, radical transformation of the landscape is among the most serious environmental issues. Erosion, land degradation, and disappearing wildlife resources are common forms of destruction in overpopulated subtropical and tropical countries.

Central and South America

In Central and South America, including the West Indies, the main problem is again the rapid population growth that causes an accelerating pressure on various types of land—even those that were previously little affected. The most rapidly growing region in the world is Central America. If the 1967 rate of growth continues, the population will double in about twenty years. The world's highest growth rates are in Colombia (4 percent), Costa Rica (3.8 percent), El Salvador (3.7 percent), the Dominican Republic (3.6 percent), and Venezuela (3.6 percent). A direct result of the developing population crisis is heavy environmental destruction leading to economic and social difficulties. Forests vanish quickly, and erosion takes over, killing previously fertile land.

In addition, the renewable natural resources throughout the South American continent are damaged at an increasing rate by gross land misuse. Along the rivers of the Amazon basin the forests are destroyed, and valuable freshwater and land wildlife is exploited through unrestricted hunting to the point of extermination. Little concern for conservation is shown by governments and people, despite the devoted work of a few conservationists. In general, there is no public response to their efforts. Competent biologists are seldom or never represented at levels where decision-makers discuss development plans for the continent's future.

South America contains one of the richest faunas of the world. Both birds and mammals exhibit a pronounced diversity in response to a wide range of habitats. Wise utilization could develop wildlife in South America into a formidable resource, but instead the wild animals are subject to serious overexploitation. This plundering goes on without regard for the fact that the wildlife capital is shrinking and will soon be exhausted. The overexploitation of wildlife seems to be traditionally embedded in the habitats and customs of most rural populations.

An interesting study of the economic value of Ecuadorian wildlife has been made by Richard E. Warner (1968). The economic picture of wildlife exploitation in Ecuador includes three principal components. Conservative estimates of annual productivity values are:

1. Wildlife used as food by the population:	$1,515,000
2. Wildlife products (mostly hides) sold commercially:	1,280,350
3. Wildlife exported alive (to zoos and for pets):	250,000

The current overall economic value of Ecuadorian wildlife is therefore estimated to be $3,045,000. These figures are based on low estimates. With more precise data the figures would be revised upward. Moreover, Ecuadorian wildlife is depleted. With sound management measures its normal potential could be reestablished, and productivity increased many times. In addition, some aspects of the economic picture have been omitted entirely—for example, the value of sport hunting, recreation, and tourism. The entire wildlife exploitation activity of the Galapagos Islands inhabitants has been excluded.

These facts show that wildlife as a natural resource occupies a place of far greater significance in the economy of the country than has hitherto been realized. It also indicates the critical importance of establishing, as soon as possible, a conservation and management program so that this vitally important resource can be protected from destruction by overexploitation, and managed to provide long-term benefits for the human population. This example from the relatively small country of Ecuador is not peculiar to that nation. It is representative of most areas of South America, and clearly shows the need to take immediate action to preserve, manage, and utilize the continent's wildlife resources in a productive way.

Europe

Densely populated and highly industrialized Europe is confronted with those problems that are specific for developed countries: pollution of air, water, and soil, including environmental poisoning by biocides. Pesticides in the form of complex chemical compounds are still used in large quantities and almost indiscriminately in many countries, causing incalculable damage to the environment, particularly fish and wildlife resources, and exposing man to great risks.

The largest part of Europe is the U.S.S.R. The hard exploitation of natural resources in that country has virtually turned ecosystems upside down, with dangerous consequences to long-term productivity. The redirecting of rivers that originally flowed northward to the Arctic Ocean, so that they now are parts of the Volga, Dniepr, and Don River systems, has brought about a series of disturbances and fatal effects, such as a tremendous decrease in natural productivity, change of climate, creation of swamps, salinization, erosion, deterioration of soils, population decreases of important fish and wildlife, all causing heavy losses to industry, agriculture, fisheries, and hunting.

Once central Europe was covered with vast deciduous forests extending from the Ural Mountains to the Atlantic. Very little remains of this natural setting. Thus far the disappearance of deciduous forest from Europe has not proved catastrophic to man; on the contrary, the results have been advantageous. The cultivated land has supported a growing population for centuries without, miraculously, much impairing the fertility of the soil. A comparison with the Mediterranean zone emphasizes the fertility of central Europe, due mainly to the latter's temperate climate. However, many signs indicate that the productive life of the European plain is slowly being exhausted through misuse. Spring floods, which once were distributed over the countryside by an endless network of rivers and streams, and which at times flooded the meadows and were stored in marshes or in the soil, have been canalized, and the soil is no longer enriched by them. The groundwater level has sunk, in many places catastrophically, and desiccation has set in.

Mediterranean Europe is suffering from too heavy deforestation, and millennia of overgrazing that coupled with recent population pressure, does not allow land recovery. As one flies eastward over the Mediterranean countries from the Iberian peninsula, one sees everywhere—on the large islands in the sea, in the ancient lands of Italy, in the mountains and archipelagos of Greece—eroded areas like open wounds in the earth, and one realizes with a shock the incredible extent to which man has destroyed his environment. As one passes above Portugal and Spain, it is already evident that vast expanses have become a desert—dead ground. As this reality sinks home, one is reminded that the Iberian soil has almost the same ancient history as that of Greece and Italy. Phoenicians, Carthaginians, Romans came and vanished, all exploiting the natural resources in ruthless and uneconomic ways, exhausting the forests and fertile soils, leaving the earth unprotected from winds and winter rains. Later, the Moorish occupation of Iberia made its contribution to the impoverishment of arable lands. The result is that Portugal and Spain are perhaps even more eroded than Italy and Greece.

Over the centuries, most of the topsoil has been washed away, and what was left on the mountains has been blown down into the valleys. There, for the time being, it has taken the place of the original soil, which washed away because of the destruction of the vegetation. This flight of soil is still going on, but the reserves in the highlands will soon be exhausted, and the valleys will then no longer be replenished from the heights. This will mean the end of agriculture in the valleys and on the plains of the lowlands, unless there is radical reforestation.

One sees the same depressing scene over the large Mediterranean islands and over Italy and Greece. Some green oases remain, either vestiges from earlier times or the results of persevering toil in the face of adverse conditions. Here and there a valley even seems rich with flourishing vegetation. They inspire hope, but unfortunately they must all too often be regarded as the last manifestations of life before death sets in.

The value of the few remaining wilderness areas in Europe is immense, and should not be underestimated by the governments that still possess them. These governments have a responsibility toward the whole of Europe, where the growing urban populations turn progressively to wild areas for vacations and recreation. With a rising standard of living and cheap transportation, wilderness areas like those in Scandinavia and the U.S.S.R. will be visited increasingly by people from all directions, even from so far off as the Mediterranean countries.

Asia

As a whole, the conservation picture for Asia is highly unsatisfactory. In Siberia tremendous hydroelectric schemes threaten the function of entire ecosystems that may have detrimental long-term effects on the environment. In the temperate and cold deserts, the subdeserts, the steppes, and the high plateaus of Central Asia, incredibly rich resources have been destroyed through overgrazing by livestock and overhunting. Even so, at the end of the nineteenth century, before the spread of firearms to isolated regions and the opening up of remote areas by railroads, the central Asian plains were the home of large herds of wild asses, wild horses, gazelles of several species, saiga and Tibetan antelopes, wild sheep, yaks, and camels. The countless ungulate herds of a century ago have vanished and with them a rich protein and hide resource. These animals were remarkably well adapted to a harsh environment, constituting an optimal productivity of those areas.

The story is the same for the deserts and semideserts of southwest Asia, where wasteful land use and massacres of gazelles and antelopes have destroyed productive resources.

In India too many people and too much livestock create a number of problems—encroachment of deserts on previously fertile lands due to overgrazing and agriculture, the decline of productive habitats, the disappearance of one of the world's richest and most spectacular faunas containing a high number of large herbivorous animals and representing protein resources of great value.

In China the far-reaching damage of the past, gigantic forest destructions, and severe population pressure, with few attempts at environmental restoration measures, cause continued land deterioration, producing a dangerous situation and crisis similar to that in India.

Southeast Asia, like India and China, is confronted with tremendous population growth and rapid destruction of renewable natural resources. Thanks to volcanic activities in Indonesia, ashes fertilize the soil and delay the negative consequences of soil deterioration, which seems to cause little concern for undertaking an effective conservation program. The ambitious Mekong River development project involves technological teams representing twenty-eight countries engaged in construction of an enormous water-control system involving 236,000 square miles of the lower Mekong basin, where more than thirty million people live. Beyond its immediate economic advantages, this project may well create long-term negative effects of an unforeseen magnitude simply because ecological aspects have been neglected in the planning stage. Harnessing giants in nature is a much more dangerous game than modern technology seems to realize.

Some few remaining forest savannas in the Mekong region of Cambodia still harbour ungulate mammals of great nutritional and scientific value. The same is true in Thailand and Burma, but these wild patches vanish quickly due to habitat destruction and overhunting.

The Philippines are developing rapidly, but seen in long-term perspective they are depleting their resources more quickly than other Asian countries. In particular, the forests are being destroyed at an extraordinarily rapid rate due chiefly to shifting cultivation. The latter form of cultivation is—in addition to the population increase of which it is a consequence—one of the most serious environmental problems in all of southeastern Asia.

Africa

Africa's principal conservation problem is proper land use. There is much evidence that the wild plant and animal resources of Africa are of great economic importance for protein and hide production and for the tourism industry, in addition to being of pronounced cultural and scientific value. Conservation, management, and utilization of animal resources are not difficult, but it is not so easy to reorient people from their traditional cultural habits. Therefore conservation of wildlife in Africa, as in Latin America and Asia, is more concerned with people than with animals.

Yet in many parts of tropical Africa wild animals provide most of the animal protein. The abundance of large mammals on the semiarid savannas of Africa is unmatched elsewhere in the world. Despite this obvious fact, the animal resources are plundered or deliberately eradicated to be replaced with livestock. The domesticated animals rapidly kill the environment before they are forced to abandon the land due to their own destruction. To let the actual overgrazing by livestock in Africa go on is a policy of suicide. It is true, however, that the habitat deterioration of pastoralists is less serious than that caused by agriculturists.

Another feature of Africa's wild animals is their inestimable value as a world asset. The African countries are custodians of a unique world heritage. Fortunately, in several countries—up to now chiefly in East Africa—this heritage has proved to be a prime income-producing industry. It is essential to keep the remaining wild areas intact, because they may well be the only ones in Africa man will ever have. The realization of the tourist potential of national parks and nature reserves, together with an appreciation of the economic returns, should encourage the establishment of like conservation areas on a much greater scale all over the continent.

Africa still has some time to solve its problems before its population growth has reached proportions comparable with those in Asia and South America. The continued destruction of forests, particularly mountain forests, has far-reaching negative environmental effects that constitute one of Africa's most serious problems. The deforestation has various detrimental consequences for the landscape. It jeopardizes maintenance of the water balance, productivity of soils, the stability of "environmental health," and even local climates.

An overriding factor in the need to preserve the African forests is the importance of forest cover for its water-regulation properties —in short, for its general ecological role. Almost all montane rain forests in Africa are threatened by a rapidly increasing commercial utilization of their timber, by shifting cultivation and by deforestation for pastures. At present few of them remain in virgin condition. Many development projects in Africa, often suggested by the FAO and by governmental agencies in Europe and North America, transform montane forests by exploiting them for timber and replacing them with plantations of exotic conifers and eucalypts. These cultivated forests cannot compete in efficiency with natural montane forests as accumulators and distributors of surface runoff, and as producers of nutritional soils and animal proteins. Where the forests are gone, erosion and declining water resources lead to the collapse of entire regions.

It is imperative for Africa's future to plan the management of wild areas—semideserts, savannas, forest, and wetlands—and to restore those that have been destroyed by unwise land use. Such planning requires basic ecological research to gain an understanding of the complexity of habitats in each area concerned.

Australasia

Australia, with New Guinea and New Zealand, is the only continent except Antarctica that does not have a population problem. In Australia the political ambition is simply to fill up the empty quarters with people through immigration.

Compared with Africa and Asia, man entered Australia relatively late, probably less than 20,000 years ago. Only during the past 180 years, however, have human beings had any far-reaching environmental effects there. The story of the tremendous destruction of habitats and animal life after the arrival of the Europeans is well known among biologists. Twenty-six species of mammals and thirteen species of birds have been exterminated in Australia during recent times. What is not so well known is that nature destruction, rape of forests, and depletion of animal populations are still proceeding on a major scale.

The first European colonists reached Australia in 1788. It is not clear what influence early man had on the island continent. Many research workers hold the view that in spite of more than 10,000 years of occupation, the aborigines with their stone tools had little effect on nature. Others believe that grass fires that were periodically lit to catch lizards—a practice that still continues—contributed greatly to environmental changes. As Dr. Paul S. Martin has pointed out, prehistoric man should not be dismissed as a possible plunderer of Australia. If it develops that the earliest Australians had something to do with the extinction of *Diprotodon, Nototherium, Palorchestes, Sthenurus,* and a large number of other Pleistocene giant marsupials and birds, the reprehensible record of extinctions by the European settlers will look like small potatoes.

Regardless of which view of the ancient past is correct, the recent record is clear. The arrival of Europeans, with sheep, cows, and rabbits in their wake, led to a collapse of ancient ecosystems. By 1890 there were 100 million sheep in Australia. Today the total is at 160 million. As human prosperity increased, few noticed that the sheep were trampling the wildlife to extinction. Because they were accused of competing with the sheep, the larger kangaroos were killed in enormous numbers. They are still being killed in large quantities. In New South Wales alone, 66 percent of the red kangaroo (*Macropus rufus*) population has apparently been killed since 1963. In some areas in western New South Wales and central Australia a combination of drought and killing has virtually wiped them out. Professional exterminators in Australia are estimated to have shot about 2 million kangaroos in two years. The red kangaroo could well be on the road to extinction.

There are no national parks in Australia where this species could be protected. The existing faunal sanctuaries have little permanent status. In addition, grazing by cattle or sheep on the outskirts, and the lack of policing by wardens, make the reserves a gesture rather than an accomplished fact. In New South Wales, for example, one percent of the state is a nature reserve, three-quarters of this the Kosciusko State Park in the Australian Alps. Conservation of nature in Australia is not effective because commonwealth legislation in this field is lacking, and the laws of the states are quite heterogeneous.

The fifty-five tons of kangaroo meat exported weekly represent 10,000 kangaroos (*M. rufus* and *M. canguru*) and wallaroos (*M. robustus*) killed. Many species of the smaller wallabies are also heavily hunted. There is nothing wrong with cropping an animal for meat or fur, but this does not mean exterminating it. It is claimed that these reductions are necessary to keep the sheep-grazing areas free from damaging kangaroos. The old story about one kangaroo eating as much as eight sheep is still widespread. Many Australian biologists think that kangaroo farming might be more profitable than sheep farming. The sustained yield would be higher and there would be no destruction of vegetation and soil.

Dr. H. J. Frith, director of the Division of Wildlife Research of the Commonwealth Scientific and Industrial Research Organization (CSIRO) confirms that kangaroos are the most productive animals of Australia's arid lands. He states that kangaroos must be used, and sheep and cattle numbers reduced, if the productivity of Australia's arid and semiarid grazing lands is to be conserved. Kangaroos are much more efficient than sheep at converting vegetable food to animal protein, and their carcasses provide a considerable higher proportion of edible protein in the form of lean muscles. Moreover, kangaroos are much better adapted to the inland environment than are livestock—for example, in their ability to withstand drought. This is like Asian and African ungulates in similar habitats.

An urgent need exists to adopt a commonwealth program of conservation, management, and controlled utilization of the kangaroos as a productive and habitat-conserving natural resource. Such a program should be part of a general pasture-management program in Australia.

As mentioned above Australia needs more national parks and nature reserves. The states have been slow to protect the wide range of unique habitats and ecosystems adequately, and the commonwealth government has not acted at all. In recent years, however, a change for the better has taken place in Australian attitudes toward national parks and conservation in general, mostly due to the magnificent work of the recently established Australian Conservation Foundation. The present disparity existing between the various states as regards a national conservation policy and leadership makes a federal conservation agency most desirable, not to say necessary. The CSIRO is already at federal level carrying out conservation research which can serve as guidelines for conservation action on a continental scale.

Australia has a great obligation to the world to conserve the Great Barrier Reef, the largest structure ever created by living organisms. Oil prospecting, mining activities, and other exploitations threaten the living coral reefs off Australia. Fortunately, these threats seem to have excited public interest more than any other conservation issue in Australia. This reaction is promising.

In New Zealand the European newcomers have likewise in a short time succeeded in drastically changing the natural environment by introductions of exotic grazers and browsers, and by deforestation. Geologically young New Zealand, with its rugged topography, thin soil, semitropical forests, and high rainfall, is sensitive to such environmental changes. The previously dense forests have altered profoundly in appearance as the result of exotic trees being planted. They have also diminished in acreage, with a consequent, pronounced increase in erosion rate and decline in the ability of watersheds to retain water.

Oceania

Almost all islands with human populations have been subjected to widespread destruction of vegetation and animals. This deterioration goes on and in many areas has caused irreversible destruction. The few islands that remain with virgin vegetation and intact animal life, or that have not been too markedly damaged by human activity, are of inestimable scientific value and may well prove significant for human welfare.

Antarctica

The continent of Antarctica is almost completely covered with an ice cap that hides plains and mountains. Actually it may be an immense archipelago. Nevertheless, this continent of glacial ice is already marked by man in many ways. The Antarctic penguins and seals are contaminated with DDT residues, probably through the marine food chains.

A conservation treaty for Antarctica exists among the nations that are active on this continent and share the responsibility for it. The treaty is needed, for Antarctica will probably be the last wild refuge of planet Earth.

The Oceans

Some 70 percent of the globe consists of oceans. They are often looked upon by man as his ultimate food resource, which hopefully could provide for the human population increase. Yet we have not learned from the large whales the lesson of how to use plankton organisms directly in an economic way instead of utilizing the final links in the marine food chain. This shortcoming is probably an advantage because if man had been technically able to catch and utilize plankton organisms in unlimited quantities, this resource certainly would long since have been reduced to such an extent that the marine ecosystem, too, would be threatened.

The greatest conservation problem of the oceans is that they are "no man's sea." No national legislation controls the exploitation. A number of countries exploit marine resources without consideration for future needs or the fact that they are plundering a global property. These countries would not treat the marine resources within their own territorial waters in the same unwise way. Selfish greediness characterizes this modern form of piracy in international waters. The situation has led to continuous overexploitation of many marine animal resources: invertebrates, fish, seals, whales. Some international agreements exist, but they cover no more than a fraction of what should be regulated in the interest of mankind.

The most urgent problem is the whaling industry, of course, and man is hunting these marine animals to the point of extinction—especially the larger, slow-breeding whales. For this reason alone it is imperative that the whaling nations abstain from catching threatened species.

Stocks of commercial species of fish that are under intense fishing pressure may be depleted all over their range. Therefore, marine nature reserves should be established along the continental shelves in order to provide refuges for breeding populations or younger classes from which replenishment of the fish populations can be made. The establishment of marine national parks should be encouraged, particularly for the protection of coral reefs.

CHAPTER ELEVEN

The Future

Without careful planning and world population control, mankind's future will be in grave jeopardy. Without restrictions on human numbers, there will be either wars or a drift into constant tensions, riots, and social unrest, until environmental resistance barriers in one form or another drastically reduce the population level. Without stopping the growth of human population and replacing it with a population decline, there is no chance to solve the main environmental problems which on an increasingly alarming scale are facing man. The question is whether it may not already be too late. It is possible that ways will be found to delay population crashes—for example, through the agency of a totalitarian system in which individuals are collectivized, or by some kind of rigid social organization resembling that of ants and termites. Such automatization of people would mean a loss of intellectual freedom, to which normal human beings would have difficulty adapting. Man's need for a minimum of living space will be denied, probably resulting in mass psychotic behaviour. This in itself would comprise an environmental resistance barrier.

These prospects make for ugly prophecy full of pessimism. Yet there are reasons to remain optimistic because man, through intelligent action and responsible cooperation, can avoid the ultimate fatal consequences of the population and environmental crisis. That crisis is already with us, and it has tremendous dimensions. In fact, it is without precedence in human existence. Man's principal task as a global citizen is to overcome this crisis before it develops into a catastrophe. The hitherto laissez-faire attitude in the industrial countries has been due partly to the economic concept of an ever-continuing expansion, thought to be intimately coupled with the maintenance of prosperity. It is a tragic irony of our time that environmental deterioration problems have usually been fostered by actions intended to improve the human standard of living. Mass starvation is a consequence of public health measures and medical progress, environmental pollution is a consequence of economic growth. But people have begun to realize that ever-growing industrial development is in itself a limitation—even a threat to progress—if environmental qualities are regarded as essential parts of human life.

The Living Landscape

The quality of the living landscape around us reflects the environmental health and the future possibilities for survival. To most people the landscape they see is characterized by its vegetation. In reality it consists of much more than that. It is a community of living things—plants, trees, animals, including ourselves. Man is an integral part of this environment—not just a spectator viewing it. His impact on the environment is so great as to constitute a formidable challenge to himself. His welfare depends on using the living landscape in a way that causes a minimum of deterioration, despite growing populations and expanding exploitation pressures.

Man's dilemma is that the world's natural resources will be wasted away if he fails to comprehend the ecology of the living landscape and act accordingly. This problem is of such magnitude and global importance for the future of mankind that it demands intergovernmental involvement. Conservation principles must be implemented on an international scale. Conservation is applied ecology. Actions taken must be based primarily on ecological concepts. It is also true that the environmental quality will deteriorate as long as the present population growth continues unchecked. Humanity, surrounded by a polluted environment, is at present facing an extremely dangerous situation, from which only a quick but tremendously costly action may allow us to escape.

Food for Tomorrow and the "Green Revolution"

Hitherto world food production has not been able to cope with the population increase. Yet many agricultural experts claim that the food yield can be increased many times, and can even be made to keep pace with the expanding population. The director-general of

FAO stated at the UNESCO Conference on the Biosphere in Paris in September, 1968, that FAO has no statistical evidence suggesting that the world is incapable of feeding its rising population. He saw no "justification for neo-Malthusian predictions of disaster."

Many biologists cannot agree with this optimism. They are skeptical about improvement in world-wide food production because even today at least two-thirds of the world population does not have enough to eat. Many feel that the battle to feed the world population is already lost. They foresee that by 1985 there will be world-wide famines in which hundreds of millions of people will starve to death. They predict that these catastrophic famines will lead to revolution, social turmoil, and economic upheaval sweeping areas of Asia, Africa, and Latin America.

The world has a greater potential of food production than its present yield. Enormous areas with high natural protein productivity have been destroyed or are used in detrimental ways, particularly in subtropical and tropical countries. Land restoration can make it possible to increase food production through cultivation and cropping of wildlife on a sustained yield basis.

It is true that agricultural production in critical areas of Asia has recently made great progress. The Philippines have achieved self-sufficiency in rice for the first time in history in 1970 despite a galloping population increase. Malaysia and South Vietnam are predicting the same for 1971 and Indonesia for 1972. Similar trends are reported from Ceylon and Thailand. Pakistan claims it will shortly be self-sufficient in all cereals. West Pakistan is at present a net food grain exporter, and India became self-sufficient in 1971 with a food grain production of nearly 108 million tons. This is the so-called Green Revolution in Asia.

Moreover, there are marine areas that have scarcely been exploited at all. The potential, practicable production of human food and animal feed from the living resources of the oceans by current fishing techniques has been variously estimated at four or even ten times the current annual yield of fish and shellfish. We are today harvesting the oceans by hunting, not by cultivation, in contrast to what we are doing on land. Could not mariculture or marine aquaculture, involving mollusks, crustaceans, and fish, be a zoologic parallel to botanic agriculture? Harold H. Webber (1968) has discussed this possibility and come to the conclusion that the immediate return from mariculture will probably contribute very little to relief from hunger for the undernourished peoples of the world. It is even unlikely that the caloric requirements of the hungry peoples can ever be met from the sea.

However, several other ways to increase marine harvests are possible. A number of potential food resources both in the plant and animal world, in the seas and on land, have not yet been rationally utilized. The question is whether these resources can be exploited without being destroyed, in order to catch up with the food needs of the world's growing population. The necessary annual increase for doing this is estimated at 2.25 percent in the next fifteen years. But what will happen thereafter?

The 1970 Nobel Peace Prize winner, Norman Borlaug, claimed, in 1965, that "man can feed the world's mushrooming human population for the next 100–200 years." Four years later Borlaug shortened the time to "two or three decades." The biologists' skepticism about the agriculturists' proclamations of "miracle grains" and their potential for feeding increasing populations of man for centuries or decades ahead seems well founded. The agricultural dream, or rather dogmatism, to eradicate hunger from our world does not seem to take ecological realities into account. Many agriculturists think the Green Revolution will overcome the earth's biological limitations. It is an extremely dangerous attitude, an illusion that leads to a slowdown in the efforts to save man and his environment through ecological planning, which obviously must include population control.

The high-yielding varieties of wheat and rice resulting from experiments in Mexico and the Philippines are the basis for the Green Revolution, which also involves the use of fertilizers, pesticides, irrigation schemes, the opening of new lands for monocultures, and massive financial subsidies. Initially, agricultural production increased in southern Asia as a result of the Green Revolution, but the price was a general deterioration of other natural resources which in the long run are essential for keeping the environment in good health. In addition, economically the technology behind the Green Revolution is so costly that few farmers can afford to use it without getting economic support. In other words, the Green Revolution seems to ignore the real costs of wheat and rice production. The new rice varieties require costly irrigated and carefully controlled lands. On nonirrigated lands these new varieties do no better than the old ones. The sudden surplus of these crops can be attributed, in part, to subsidies, price manipulation, and deliberate interventions in the market.

It is clear to most ecologists that the recent food increase in Asia, whether subsidized or not, cannot keep pace with the population growth. Any increase of food means that more people live and reproduce, accelerating the population crisis still more.

Wheat and rice provide about three-quarters of the world's total food grain supply consumed by man. The success of certain varieties of wheat and rice by producing a substantial increase, after a period of twenty years when national yields in Asia were stagnant, gives humanity more years to restrain population growth before a new food crisis develops. "All attempts to raise food production will be futile if population growth is not slowed drastically," writes D. S. Athwal (1971) of the International Rice Research Institute in the Philippines. If the present population growth rate continues, the world must double the quantity of food it is now producing.

The initial success of the Green Revolution is partly explained by unexpectedly favourable weather conditions, which greatly improved the yield per acre in Asia. This important environmental factor seems to have been overlooked by those praising the miracles of the Green Revolution. As William C. Paddock, a specialist in agriculture, said at a symposium on "Man: His Environment, His Future" at North Carolina State University in Raleigh in 1970, the Green Revolution is green only because it is viewed through green-coloured glasses. Far and away the most important factor for increased agricultural production in southern Asia during recent last years was the improvement of weather.

Throughout Asia virtually all crops increased during the last years of the 1960's, even crops without new high-yielding varieties and in countries without the "miracle seeds." Weather seldom gets the credit it deserves. When crops are poor, governments put the blame on the weather. When crops are good, governments and agriculturists and aid agencies take the credit, heralding their foresight and wisdom of planning, providing fertilizers, toxic chemicals, and new crop varieties (which several years later undoubtedly will turn out to be highly vulnerable to disease due to lack of genetical diversity). The genetic changes in producing crops of miracle wheat or wonder-size varieties having similar disease-resistant characters in enormous monocultures over large areas, where they are exposed to attacks from parasites due to their loss of variability and "dynamic self-defense," cannot be overestimated, at least not by ecologists.

Among world planners in development agencies the Green Revolution has become a sort of magic phrase, to which they are constantly referring as if the world's population problems in relation to food supply will be or already are automatically solved. To ecologists and conservationists such a philosophy is irresponsible.

FAO believes that if the policies recommended in its Indicative World Plan for the 1970's and the early 1980's are fulfilled, the world's agricultural production will rise at an average rate of 3.7 percent per year between 1970 and 1985, which will exceed by 1 percent the expected increase in the world's population over this period.

Even on purely economic grounds the Green Revolution seems to be unrealistic. According to FAO's Second World Food Conference at The Hague in 1970, the food demand in developing countries will be nearly two and a half times the level of 1962, and two-thirds of this extra demand will be a consequence of population growth. The FAO plan to meet this food requirement calls for an expenditure of $112 billion, of which $37 billion alone is required for expanding irrigation (another patent remedy with dangerous environmental implications). It is taken for granted that the foreign aid of industrial nations will pay for feeding the hungry millions of the developing nations. Is this economically realistic or feasible?

Biologically and ecologically the Green Revolution endangers man and his environment. It seems to require an enormous use of agricultural chemicals, an increase of 100 percent for fertilizers and of 600 percent for pesticides. It is easy to imagine and foresee the deleterious effect of this gigantic pollution on the environment, including human beings. Moreover, the talk of unexploited tropics, to be "developed" through new methods of farming which ignores the grim ecological reality that the majority of tropical lowlands and mountains are already overexploited.

The faith in irrigation schemes is another calamity for the human future. There are so many examples from tropical and subtropical lands showing disastrous failures where far more land has turned into desert than has been reclaimed through irrigation.

Or, to quote Paul Ehrlich: "Those clowns who are talking of feeding a big population in the year 2000 from make-believe 'green revolutions' . . . should learn some elementary biology, meteorology, agricultural economics and anthropology."

It is obvious that food production cannot be increased indefinitely to satisfy the needs of a continuously growing population. The problem is not one of insufficiency of food but of overreproduction of human beings. Starvation cannot be eliminated without regulating human populations through birth control.

If there is ever to be a Green Revolution on our earth it must inevitably come after a solution of the population problem. It is the latter that requires a revolution in human attitudes and behaviour.

Planning of Populations

If man is to survive, he cannot let his numbers grow indefinitely. He must take control before he has consumed or destroyed the natural resources on which he bases his existence. At one time man lived in a self-regulating environment that he did not affect. Now he profoundly affects his environment, but the environment ultimately controls him. Hence, man is enmeshing himself in his own technological trap. In social terms, the evidence is clear: crowding, poverty, starvation, increase in violent behavior, decrease in respect for life. Man must accept the cruel fact that the physical resources of the earth are already insufficient to support the present world population. He is consuming the earth's capital at an accelerating pace.

"Whatever your cause, it is a lost cause if we cannot limit our population" is a very true slogan produced by the International Planned Parenthood Federation. Family planning as it has been practiced in various parts of the world during the last decade is inadequate, and absolutely not a long-term solution. It simply does not work, and workable measures are urgent. Therefore man cannot escape from his obligation to institute population control. It must be based on a careful population planning—something quite different from family planning. Ecologists, in cooperation with sociologists and economists, must take the leadership in this task. Irrational, short-term solutions are to be avoided. The goal must be a birth rate that does not exceed the death rate. But in many regions of the world one must go beyond this point and reduce the population to a level corresponding to the carrying capacity of the environment.

Even if a global population control program could be initiated immediately and implemented with success, it would take many decades before a significant slowdown of population growth would be visible, and an even longer time, presumably several centuries, before a reduction of population to optimal sizes in relation to the environment could be achieved. In the meantime, the gap between food resources and human populations will probably widen due to continued destruction of renewable natural resources, in other words, the resource capital. That is the human dilemma.

CHAPTER TWELVE

An Ecological Strategy

Man is transforming the earth on an ever larger scale, by using up its resources through tremendous technological power. What is happening in every corner of the planet is of concern to every inhabitant, to entire populations.

At the present time, unfortunately much too late, the environmental deterioration and its threat to human survival and standard of living have become one of the major political issues of the 1970's. The population increase is for the moment out of control and so is the technology. Both lead to ecological disaster through environmental degradation. Perhaps the prime cause of the technological threat to the environment is the increasing demands constantly being made by expanding populations all over the world. It is a paradox that the improvement of standards of living through technological means results in a lowering of the environmental quality, which eventually lower the standards of living.

International Responsibility and the Ecological Crisis

It has been claimed that nobody is responsible for the ecological crisis because until recently we have not been aware of the existence of such a crisis. This is a naïve statement and, moreover, it is not true. Ecologists and conservationists have for decades been warning of the severe ecological risks of human misuse of natural resources, but few people have listened or understood.

Throughout the history of man, whatever century or continent of the world one analyzes, it is quite clear that ecological ignorance, tremendous carelessness, lack of foresight, and unbelievable human greed have led to the plundering of the earth. Modern technology has accelerated the destruction.

To save mankind from catastrophe, an ecological strategy must be developed to deal with environmental problems at their roots and in a way that everybody understands. It must necessarily be an international responsibility to formulate a policy and outline a strategy for effective ecological action on a global scale. Furthermore the people of the world must be educated to understand why in the long run an ecologically based strategy will restore the environment of our world and the confidence for the future in human minds.

We are for the moment living in an evolutionary age, and it is possible that profound social changes will occur. The youth revolt is an expression of this social and ecological unrest. The family as the basic unit of society is disintegrating everywhere, among rich and poor. The social tension is at least partly released by fear for the environmental future and for increasing urban schemes. How present societies respond to these social changes may be crucial not only for themselves but also for future human existence.

Ecology as a philosophy may well have the potential to develop into a kind of religion for younger generations of today and the world of tomorrow. It has so much realistic appeal to those who see a biological meaning in life and its evolution. With ecological awareness, conservation would automatically become an exercise in applied ecology to provide man his environmental present and future needs by maintaining qualitative habitats. The tireless activities of a small minority of foresighted conservationists have for decades kept the conservation issue alive in various countries despite public disinterest and total lack of encouragement from governments. Fortunately, many scientists have taken up the ecological crisis. Intelligent people all over the world are becoming aware of the conditions in which they are forced to live. They are beginning to wake up, and some have even reached the point of getting fed up with the air pollution, water pollution, noise pollution, food pollution, and so on. However, this reaction is not enough. We have to change mankind's arrogance and/or indifference toward nature all over the earth by introducing an ecological approach to all environmental problems.

Such an ecological message is directly linked to our possibilities to survive in a healthy environment and is of concern to everyone. When do people begin to understand the realities of existence, the impact of their everyday life on the environment, how natural systems within the biosphere function, and man's inescapable dependence on ecosystems as a part of them? Must people first realize how dangerously man is influencing his environment through misuse and overexploitation?

Fostering an understanding of the proper use of renewable natural resources cannot be done on a nationwide or global scale without education. Ecological thinking must be an integral part of school and home education. It should be integrated as a basic subject at all levels from primary schools to universities, and involving the entire educational system. The ecological message must also reach adult generations through other institutions and organizations such as political parties and churches, as well as through television, radio, magazines, newspapers, and books. The matter requires urgency. Mankind cannot wait until present school generations have grown up to take "ecological action." International agencies, such as UNESCO, should establish an ecological educational campaign as their prime task.

A worldwide educational campaign should be the first great challenge in the ecological strategy. Its message is very simple because it is appealing to basic motives of individual self-interest. Independent of racial, sociological, religious, and political barriers, the "ecological strategy" involves every individual on biological grounds. Ecologically informed individuals will have strong reactions to unwise exploitation and misuse of the environment. People who understand the basic principles of ecology will draw the right conclusions from what they see around them. They will realize the unnecessary degradation of environmental qualities and, therefore, the threat to their own lives.

In fact, some governments—particularly the United States—are already aware of the seriousness of the environmental problems but need more support from society to take effective action and establish a national policy based on ecological research and concepts. Governments will sooner or later be forced by the environment to implement ecologically based policies on an international scale.

Great progress has been made during the last decade—again particularly in the United States, whose National Environmental Policy Act, signed in 1970, is promising sound decisions in environmental policy—but the conservation actions are too slow and still meet too many obstacles from society itself. In the late 1960's we have seen politicians and other influential people become active in public meetings and the press speak out about the necessity of conservation as a remedy to the ecological crisis. Conservation has become a political, popular, and newsworthy subject because it makes sense to people. Informed politicians are using conservation arguments objectively and effectively.

As a conservationist, it is gratifying to see national and international attention focussed increasingly on the environmental crisis. What the conservationists have been talking about for years is gradually being accepted. Their efforts have not been entirely in vain.

Man's history on earth has made clear the need for conservation of nature to assure human survival. Therefore a world policy of management and utilization of renewable natural resources, based on ecological considerations, must be adopted. The governments must implement this policy. But who will define it? It is desirable that the United Nations with its specialized agencies, particularly UNESCO and FAO, in cooperation with IUCN form a council of environmental management, composed of an ecological elite with global vision representing different disciplines and regions of the world. The role of this scientific body should be to assist and advise all member countries of the United Nations on the utilization of renewable natural resources, population problems, and environmental research.

Establishment of such an advisory body as the highest international scientific authority on the conservation of nature and environmental problems should be an international responsibility and obligation. Each country should loyally follow its counsel, an undertaking that in itself requires a high level of ecological knowledge and social responsibility.

Perhaps the conservation of natural resources and the environment will emerge within the United Nations as an issue of such global importance that it will inject new life into the organization and a better understanding among the nations which share the finite ecosystem called earth.

GLOSSARY

Abiotic Physical and chemical non-living (inorganic) parts of the environment.

Acacia Vernacular and scientific name for a genus of trees and shrubs representing several hundreds of species.

Aerobic Capable of living only in presence of free oxygen.

Algae Cryptogamic plants living in water and the air.

Ambergris A morbid secretion in the intestines of the sperm whale, which is used in perfumery.

Anadromous A word describing fish which spend parts of their life-cycle in the sea but reproduce in fresh water.

Anaerobic Capable of living in the absence of free oxygen.

Aquaculture Cultivation of marine vegetation as a parallel to agriculture.

Arthropod Taxonomic group of invertebrate animals including *Crustacea, Myriopoda, Insecta,* and *Arachnoidea.*

Atmosphere A gaseous structure of air surrounding the earth and composed of three sectors: the troposphere (nearest the earth), the stratosphere, and the ionosphere.

Baleen Horny plates attached to upper jaw of whalebone (baleen) whales (*Mysticeti*).

Biomass Total weight of organisms per unit area.

Biome A major life zone consisting of a complex of habitats and communities; for example, tundra, forest, desert.

Biosphere The part of the globe (air, water, and soil) containing living organisms.

Biota The flora and fauna of a region.

Biotic Biological parts (plant and/or animals) of the environment. For example: biotic community = a community of organic life or of plants and animals as a whole; biotic factors = biological factors.

Calcarious Soil derived from decomposition of limestone rocks.

Campos Drier savannas and bushlands of various types (montes, caatingas) in South America.

Carapace A chitinous or bony shield covering all or part of back of certain animals; for example, turtles and tortoises.

Carnivore Mammal belonging to the order *Carnivora* = carnivores (252 species).

Cetacean Mammal belonging to the order *Cetacea* = whales (92 species).

Chaparral Mediterranean type of vegetation (macchia, maquis, fynbos).

Copepod Taxonomic group of crustaceans, usually living as a part of freshwater and marine plankton.

Deciduous Type of foliferous forest, in which the leaves of the trees fall seasonally at the end of growth period or at maturity.

Detritus Disintegrating or decayed material of plants and animals. May also refer to other material eroded or washed away.

Ecosystem Ecological system formed by interactions between and of plants and animals with their physical and chemical environment.

Edaphic Conditions influenced by the type of soil.

Entropy One of the quantitative elements determining the thermodynamic condition of a portion of matter.

Eutrophic Rich in nutrition, well-nourished. Usually refers to freshwater bodies having abundant vegetation.

Gallinaceous Bird belonging to the order *Galliformes* = megapods, curassows, guans, chachalacas, grouse, pheasants, partridges, francolins, quails, peafowl, guineafowl, turkeys, hoatzins and others.

Hecatomb Mass quantity of certain objects.

Intraspecific Within a species. For example, intraspecific competition means competition between members of the same species.

Kilocalorie Measure of nutritional values in relation to heat production.

Littoral Zone at or near the seashore, also zone between high- and low-water marks.

Llanos Tropical or subtropical grasslands of chiefly two types in South America.

Loess Air-deposited collections of soil.

Mangrove Tropical trees and shrubs growing in the mud on the seashore down to low-water mark, with large masses of interlacing roots above ground, which intercept mud and weeds, and thus cause the land to encroach on the sea. An important habitat for many organisms.

Microclimate Temperature, humidity, luminosity, and other meteorological features characterizing the conditions in a restricted place; for example, under the snow, or a stone, on a tree trunk, a leaf or a grass stem, at the bottom of a stream, etc.

Monoculture Artificial growths of uniform crops or forests consisting of one species only.

Montane Region or organisms located in the mountains (= mountainous).

Mustelid Mammal belonging to the family *Mustelidae* = weasels, martens, wolverine, ratel, badgers, skunks, and others (68 species).

Niche The biotic and abiotic part of a habitat that provides the essential needs for a species and, therefore, is frequently or constantly utilized by it.

Oligotrophic Poor in nutrition. Usually refers to freshwater bodies with little vegetation.

Pampas Treeless grasslands in Argentina.

Plankton Marine and freshwater plants and animals of small size drifting with surrounding water or having weak locomotory power, particularly for vertical movements. Include eggs and larvae of fish.

Planctonic Wandering or drifting in water. Adjective of plankton. For example, planctonic organisms.

Phenology The science of periodic biological events (such as flowering, breeding, migration, hibernation, and so on) in relation to seasonal and other climatic factors.

Pinniped Mammal belonging to the suborder *Pinnipedia* = seals (32 species).

Prairie Treeless grassland in North America.

Primate Terrestrial and arboreal mammals belonging to the order *Primates* (tree-shrews, lemurs, aye-aye, lorisids, tarsiers, monkeys, apes, and man; 193 species).

Race See Subspecies.

Rainforest Evergreen forest with high number of tree species requiring high rainfall, high and relatively constant humidity, and usually high temperature.

Savanna Subtropical or tropical grassland of various types.

Sierra Mountain range.

Sirenian Aquatic mammal belonging to the order *Sirenia* (dugongs and manatees; 4 species).

Species Taxonomic category of plant or animal consisting of groups of interbreeding natural populations which are reproductively isolated from other such groups. A species may be divided into subspecies (races).

Steppe Xerophilous and generally treeless grassland.

Subspecies (= Race) A geographically defined aggregate of local populations which differs taxonomically from other such subdivisions of the species.

Taiga Northern coniferous forest zone. Especially used for Siberian coniferous forests, but includes the whole Eurasian coniferous forest zone.

Talus Mountain slopes often with accumulations of fallen rocks.

Temperate Climatic region located between the arctic and subtropical belts in both the northern and southern hemispheres.

Tropical Regions and organisms in the tropics, that is, the climatic belt located around the equator and between the Tropics of Cancer and Capricorn.

Tundra In strict sense a region with permanently frozen subsoil, but often used as a term for arctic, treeless regions above the timberline.

Ungulate Odd-toed (*Perissodactyla,* 16 species) and even-toed (*Artiodactyla,* 194 species) hoofed animals (*Ungulata*).

Veld Type of grassland savanna in southern Africa.

Wadi Riverbed or valley in deserts or semideserts that only occasionally is filled with water.

SELECT BIBLIOGRAPHY

At the request of a U.S. publisher, this book was written in 1969 and kept up to date until publication in 1972. A number of technical papers have been used by the author as background material, but references to all these publications have only been made in a few cases as the publisher felt such information was not necessary in a popular book. However, the major books consulted during the writing of this work are included in the bibliography. In addition, a selection of the vast number of books on conservation topics published after the completion of this book, have been listed in order to orientate the reader among useful conservation literature.

Adams, A. B., ed., *The First World Conference on National Parks.* Washington, 1964.

Allee, W. C., and Schmidt, K. P., *Ecological Animal Geography,* 2nd ed. New York, 1951.

Allee, W. C., and others, *Principles of Animal Ecology.* Philadelphia: W. B. Saunders, 1949; London: W. B. Saunders, 1950.

Allen, D., *Our Wildlife Legacy.* New York: Funk & Wagnalls, 1954; London: Mayflower, 1954; Toronto: Ryerson, 1954.

Allen, S. W., *Conserving Natural Resources: Principles and Practice in a Democracy.* New York: McGraw-Hill, 1955.

Andrewartha, H. G., and Birch, L. C., *The Distribution and Abundance of Animals.* Chicago: University of Chicago Press, 1954; London: Cambridge University Press, 1955.

Ardrey, R., *The Territorial Imperative.* New York: Atheneum, 1966; London: William Collins Sons, 1967.

Arvill, R., *Man and Environment.* Harmondsworth, England: Penguin, 1967.

Bates, M., *The Prevalence of People.* New York: Charles Scribner's Sons, 1955; London: Charles Scribner's Sons, 1956.

———, *The Forest and the Sea: A Look at the Economy of Nature and the Ecology of Man.* New York: Random House, 1960; London: Museum Press, 1961.

———, *Animal Worlds.* New York: Random House, 1963; London: Thomas Nelson & Sons, 1964.

Benarde, M. A., *Our Precarious Habitat.* New York: W. W. Norton, 1970.

Bennett, H. H., *Soil Conservation.* New York: McGraw-Hill, 1939; London: McGraw-Hill, 1940.

———, *Elements of Soil Conservation.* New York and London: McGraw-Hill, 1955.

Black, J. D., *Biological Conservation.* New York: McGraw-Hill, 1954.

Borgstrom, G., *The Hungry Planet: The Modern World at the Edge of Famine*. New York: Macmillan, 1965.

Boughey, A. S., *Ecology of Populations*. New York: Macmillan, 1968.

Bourliere, F., and Verschuren, J., *Introduction à l'écologie des ongules du Parc National Albert*, I-II. Exploration du Parc National Albert. Bruxelles, 1960.

Brady, N. C., ed., *Agriculture and the Quality of Our Environment*. Washington: The American Association for the Advancement of Science, 1967.

Bresler, J., ed., *Human Ecology: Collected Readings*. Reading, Mass.: Addison-Wesley, 1966.

Brown, H., *The Challenge of Man's Future*. New York: Viking, 1954; London: Martin Secker & Warburg, 1954; Toronto: Macmillan, 1954.

Brown, L., *Africa: Natural History*. New York: Random House, 1965; London: Hamish Hamilton, 1965.

———, *Agricultural Changes in Kenya: 1945–60*. Stanford, 1968.

Buchwald, K., and Engelhardt, W., eds., *Handbuch fur Landschaftspflege und Naturschutz*. Munchen, 1968.

Carson, R., *The Sea Around Us*. New York: Oxford University Press, 1951; London: Staples Press, 1951.

———, *Silent Spring*. Boston: Houghton Mifflin, 1962.

Clements, F. E., and Shelford, L. E., *Bio-Ecology*. New York: John Wiley & Sons, 1939; London: Chapman & Hall, 1939.

Cloudsley-Thompson, J. L., *Biology of Deserts*. London: Hafner, 1954.

Colman, E. A., *Vegetation and Watershed Management*. New York: The Ronald Press, 1953.

Commoner, B., *Science and Survival*. New York: Viking, 1966; London: Victor Gollancz, 1966; Toronto: Macmillan, 1966.

Cook, R. C., *Human Fertility: The Modern Dilemma*. New York: William Sloane, 1951; London: Victor Gollancz, 1951; Toronto: George J. McLeod, 1951.

Cooley, R. A., *Alaska: A Challenge in Conservation*. Madison: University of Wisconsin Press, 1966; Ontario: Burns and MacEachern, 1966.

Cousins, N., ed., *Freedom to Breathe*. New York, 1966.

Craighead, F. C., *A Biological and Economic Evaluation of Coyote Predation*. New York, 1951.

Crisp, D. J., ed., *Grazing in Terrestrial and Marine Environments*. Philadelphia: F. A. Davis, 1964; Oxford: Blackwell Scientific Publications, 1964.

Crowe, P. K., *World Wildlife: The Last Stand*. New York: Charles Scribner's Sons, 1970.

Curry-Lindahl, K., *Ecological Studies on Mammals, Birds, Reptiles and Amphibians in the Eastern Belgian Congo*, I-II (1956–60), Vols. 42, 87. Annales du Musée Royal de Congo Belge, Sciences Zoologiques.

———, *Contribution à l'étude des vertebres terrestres en Afrique tropicale,*

I. Exploration du Parc National Albert et du Parc National de la Kagera. Bruxelles, 1961.

———, *Europe: A Natural History*. New York: Random House, 1964; London: Hamish Hamilton, 1964.

Dajoz, R., *Précis d'écologie*. Paris, 1970.

Dale, T., and Carter, V. G., *Topsoil and Civilization*. Norman, Okla.: University of Oklahoma Press, 1955.

Dansereau, P., *Biogeography, an Ecological Perspective*. New York: The Ronald Press, 1957; Montreal: Renouf, 1957.

———, ed., *Challenge for Survival: Land, Air and Water for Man in Megalopolis*. New York: Columbia University Press, 1970.

Darling, F. F., *Wild Life in an African Territory*. New York: Oxford University Press, 1960.

———, ed., *Implications of the Rising Carbon Dioxide Content of the Atmosphere*. Washington, 1963.

———, *Wilderness and Plenty*. Boston: Houghton Mifflin, 1970.

Darling, F. F., and Eichorn, N. D., *Man and Nature in the National Parks*. Washington, 1967.

Darling, F. F., and Milton, J., *Future Environments of North America*. New York: Natural History Press, 1966.

Dasmann, R. F., *Environmental Conservation*. New York: John Wiley & Sons, 1959; London: Chapman & Hall, 1959.

———, *The Last Horizon*. New York and London: Macmillan, 1963.

———, *Wildlife Biology*. New York: John Wiley & Sons, 1964.

———, *The Destruction of California*. New York: Macmillan, 1965.

———, *A Different Kind of Country*. New York: Macmillan, 1968.

Daubenmire, R. F., *Plants and Environment: A Textbook of Plant Autecology*, 2nd ed. New York: John Wiley & Sons, 1959; London: Chapman & Hall, 1959.

Detwyler, T. R., *Man's Impact on Environment*. New York: McGraw-Hill, 1971.

Dorst, J., *Avant que nature meurt*. Neuchâtel, 1965.

———, *South America and Central America: A Natural History*. New York: Random House, 1967; London: Hamish Hamilton, 1967.

Dury, G. H., *The Face of the Earth*. London, 1959.

Ebling, F. J., *Biology and Ethics*. London and New York: Academic Press, 1969.

Ehrenfeld, D. W., *Biological Conservation*. New York: Holt, Rinehart & Winston, 1970.

Ehrlich, P. R., *The Population Bomb*. New York, 1968.

Ehrlich, P. R., and Ehrlich, A. H., *Population, Resources, Environment: Issues in Human Ecology*. San Francisco: W. H. Freeman, 1970.

Ehrlich, P. R., and Harriman, R., *How to be a Survivor*. New York: Ballantine, 1971.

Elton, C., *Voles, Mice and Lemmings: Problems in Population Dynamics*. New York and London: Oxford University Press, 1942.

Elton, C., *Animal Ecology*, 3rd ed. New York: Macmillan, 1947.
———, *The Ecology of Invasions by Plants and Animals.* New York: John Wiley & Sons, 1958; London: Methuen, 1958.
———, *The Pattern of Animal Communities.* New York: John Wiley & Sons, 1966; London: Methuen, 1966.
Eyre, S. R., *Vegetation and Soils: A World Picture.* Chicago: Aldine, 1963; London: Arnold, 1963.
Forde, C. D., *Habitat, Economy and Society*, 5th ed. New York: E. P. Dutton, 1963; London: Methuen, 1963.
Fosberg, F. R., *Principal Terrestrial Ecosystems.* Honolulu, 1961.
Gabrielson, I. N., *Wildlife Conservation.* New York: Macmillan, 1941.
Glover, P. E., *A Review of Recent Knowledge on the Relationships Between the Tsetse Fly and its Vertebrate Hosts.* IUCN Publications, New Series —6: 1–84. 1965.
Golley, F. B., and Buechner, H. K., eds., *A Practical Guide to the Study of the Productivity of Large Herbivores.* London: Blackwell Scientific Publications, 1968.
Gordon, M., *Sick Cities.* Baltimore, 1965.
Graham, E. H., *Natural Principles of Land Use.* New York and London: Oxford University Press, 1944.
———, *The Land and Wildlife.* New York and London: Oxford University Press, 1947.
Graham, F., Jr., *Since Silent Spring.* Boston: Houghton Mifflin, 1970; London: Hamish Hamilton, 1970.
Haden-Guest, S., and others, *World Geography of Forest Resources.* New York: The Ronald Press, 1956; London: Arthur F. Bird, 1958.
Handler, P., ed., *Biology and the Future of Man.* New York: Oxford University Press, 1970.
Hawkes, J., *Man on Earth.* New York: Random House, 1955; London: Cresset Press, 1954; Toronto: Ambassador Books, 1954.
Helfrich, H. W., ed., *The Environmental Crisis: Man's Struggle to Live with Himself.* New Haven: Yale University Press, 1970.
Hickey, J. J., ed., *Peregrine Falcon Populations: Their Biology and Decline.* Madison: University of Wisconsin Press, 1969.
Holdren, H. P., and Ehrlich, P. R., eds., *Global Ecology.* New York, 1971.
Hopkins, B., *Forest and Savanna: An Introduction to Tropical Plant Ecology with Special Reference to West Africa.* London: William Heinemann, 1965.
Howell, F. C., and Bourliere, F., eds., *African Ecology and Human Evolution.* Chicago: Aldine, 1964; London: Methuen, 1964.
Hynes, H. B. N., *The Biology of Polluted Waters.* Liverpool: University Press of Liverpool, 1960.
Keast, A., *Australia and the Pacific Islands: A Natural History.* New York: Random House, 1966; London: Hamish Hamilton, 1966.
Keast, A., Crock, R. L., and Christian, C. S., *Biogeography and Ecology in Australia.* Den Haag, 1959.

Keith, L. B., *Wildlife's Ten-Year Cycle*. Madison: University of Wisconsin Press, 1963; Ontario: Burns & MacEachern, 1963.

Kendeigh, S. C., *Animal Ecology*. Englewood Cliffs, N. J.: Prentice-Hall, 1962.

Kuenen, P. H., *Realms of Water*. New York: John Wiley & Sons, 1956; London: Cleaver-Hume, 1955; Ontario: Burns & MacEachern, 1955.

Lack, D., *The Natural Regulation of Animal Numbers*. London and New York: Oxford University Press, 1954.

———, *Population Studies of Birds*. New York: Oxford University Press, 1966.

Leopold, A., *Game Management*. New York and London: Charles Scribner's Sons, 1933.

———, *A Sand County Almanac*. New York and London: Oxford University Press, 1949.

Leopold, A. S., and Darling, F. F., *Wildlife in Alaska*. New York: Ronald Press, 1953.

Lockley, R. M., *Man Against Nature*. London: Andre Deutsch, 1970.

Lorenz, K., *On Aggression*. New York: Harcourt, Brace and World, 1966.

Love, R. M., and Love, G. A., eds., *Ecological Crisis: Readings for Survival*. New York: Harcourt Brace Jovanovich, 1970.

Marshall, A. J., ed., *The Great Extermination: A Guide to Anglo-Australian Cupidity, Wickedness and Waste*. London and Melbourne: William Heinemann, 1966.

Mech, L. D., *The Wolves of Isle Royale*. Washington, 1966.

Mellanby, K., Pesticides and Pollution. London: William Collins Sons, 1967.

Mergen, F., ed., *Man and His Environment: The Ecological Limits of Optimism*. New Haven, 1970.

Milne, L. J., and Milne, M., *The Balance of Nature*. New York: Alfred A. Knopf, 1960; Toronto: McClelland & Stewart, 1960.

Mitchell, J. G., ed., *The Sierra Club Handbook for Environment Activists*. New York.

Murie, A., *The Wolves of Mount McKinley*. Washington: Government Printing Office, 1944.

Newbigin, M. I., *Plant and Animal Geography*. New York: E. P. Dutton, 1949.

Nicholson, M., *The Environmental Revolution: A Guide for the New Masters of the World*. New York: McGraw-Hill, 1970; London: Hodder & Stoughton, 1970.

Odum, E. P., *Fundamentals of Ecology*, 3rd ed. Philadelphia and London, 1971.

Odum, H. T., *Environment, Power, and Society*. New York: Interscience, 1970.

Osborn, F., *Our Plundered Planet*. New York: Grosset & Dunlap, 1951.

———, *The Limits of the Earth*. Boston: Little Brown, 1953; Toronto: McClelland & Stewart, 1953; London: Faber & Faber, 1954.

Osborn, F., ed., *Our Crowded Planet: Essays on the Pressures of Population.* New York: Doubleday, 1962.

Owen, O. S., *Natural Resource Conservation: An Ecological Approach.* New York: Macmillan, 1971.

Pfeffer, P., *Asia: A Natural History.* New York: Random House, 1968; London: Hamish Hamilton, 1968.

Philip, Duke of Edinburgh, and Fisher, J., *Wildlife Crisis.* New York: Cowles, 1970; London: Hamish Hamilton, 1970.

Phillipps, J., *Agriculture and Ecology in Africa: A Study of Actual and Potential Development South of the Sahara.* University Place, N.Y.: Frederick A. Praeger, 1960; London: Faber & Faber, 1959.

Puri, G. S., *Indian Forest Ecology.* New Delhi: Oxford Book and Stationery, 1960; London: George Allen & Unwin, 1962.

Revelle, R., and Landsberg, H. H., *America's Changing Environment.* Boston: Houghton Mifflin, 1970.

Revelle, R., Khosla, A., and Vinooskis, M., *The Survival Equation: Man, Resources and His Environment.* New York, 1971.

Richards, P. W., *The Tropical Rain Forest: An Ecological Study.* New York and London: Cambridge University Press, 1952.

Roosevelt, N., *Conservation: Now or Never.* New York: Dodd, Mead, 1970.

Rudd, R. L., *Pesticides and the Living Landscape.* Madison: University of Wisconsin Press, 1964; Ontario: Burns and MacEachern, 1964.

Russell, E. W., *The Overfishing Problem.* New York: Macmillan, 1942; London: Cambridge University Press, 1942.

Russell, E. W., ed., *The Natural Resources of East Africa.* Oxford: Basil Blackwell & Mott, 1963.

Ruttner, F., *Fundamentals of Limnology.* Toronto: University of Toronto Press, 1953; London: Oxford University Press, 1953.

Sampson, A. W., *Range Management: Principles and Practices.* New York: John Wiley & Sons, 1952; London: Chapman & Hall, 1952.

Sanderson, I. T., *The Natural Wonders of North America.* London: Hamish Hamilton, 1962.

Schaller, G. B., *The Deer and the Tiger: A Study of Wildlife in India.* Chicago: University of Chicago Press, 1967.

Sears, P. B., *Deserts on the March.* Norman, Okla.: University of Oklahoma Press, 1935.

Selye, H., *The Stress of Life.* New York: McGraw-Hill, 1956.

Serventy, V., *A Continent in Danger.* New York: Reynal, 1966; London: Andre Deutsch, 1966.

Shelford, V. E., *The Ecology of North America.* Urbana: University of Illinois Press, 1963.

Simpson, G. G., *Evolution and Geography.* Eugene, 1953.

Smith, G. H., *Conservation of Natural Resources.* New York: John Wiley & Sons, 1950; London: Chapman & Hall, 1951.

———, ed., *Conservation of Natural Resources,* 2nd ed. New York: John Wiley & Sons, 1958; London: Chapman & Hall, 1958.

Smith, R. L., *Ecology and Field Biology*. New York and London: Harper & Row, 1966.

Snow, D. W., *A Study of Blackbirds*. London: George Allen & Unwin, 1958.

Sondheimer, E., and Simeone, J. B., *Chemical Ecology*. New York and London: Academic Press, 1970.

Stamp, L. D., *A History of Land Use in Arid Regions*. UNESCO, Paris, 1961.

Swift, J. J., ed., *Proceeds of the First European Meeting on Wildfowl Conservation*. St. Andrews, Scotland, 1963. Nature Conservancy, London: 1964.

Talbot, L. M., *Wild Animals as a Source of Food*. U. S. Department of Interior. Special Scientific Report—Wildlife. 98 (1966):1016.

———, and others. *The Meat Production Potential of Wild Animals in Africa*. Technical Communication No. 16. Commonwealth Agricultural Bureau. Farnham Royal, U.K., 1965.

Tener, J. S., Fyfe, R., and Keith, J. A., *Pesticides and Wildlife*. Ottawa, 1971.

Thomas, W. L., Jr., ed., *Man's Role in Changing the Face of the Earth*. Chicago: University of Chicago Press, 1956; London: Cambridge University Press, 1956.

Udall, S. L., *The Quiet Crisis*. New York: Holt, Rinehart and Winston, 1963.

Van Dyne, G. M., ed., *The Ecosystem Concept in Natural Resource Management*. New York and London, 1969.

Vernberg, F. J., Vernberg, W. B., and Baruch, V. W., *The Animal and the Environment*. New York, 1970.

Vogt, W., *Road to Survival*. New York: William Sloane, 1948; London: Victor Gollancz, 1949.

Watson, A., ed., *Animal Populations in Relation to Their Food Resources*. Oxford: Blackwell Scientific Publications, 1970.

Weaver, J. E., and Clements, F. E., *Plant Ecology*, 2nd ed. New York: McGraw-Hill, 1938.

Whittaker, R. H., *Communities and Ecosystems*. New York: Macmillan, 1970.

Woodbury, A. M., *Principles of General Ecology*. Toronto: Blakiston, 1954.

Wrigley, G., *Tropical Agriculture: The Development of Production*. London: B. T. Batsford, 1961.

Wynne-Edwards, V. C., *Animal Dispersion in Relation to Social Behaviour*. London: Hafner, 1962; Toronto: Clarke, Irwin, 1962.

Working Conference on Birds of Prey and Owls. Caen, France, 1964. International Council for Bird Preservation, London.

UNESCO, 1963. *A Review of the Natural Resources of the African Continent*. Paris.

UNESCO, 1970. *Utilization and Conservation of the Biosphere*. Paris.

1963. *Conservation of Nature and Natural Resources in Modern African States.* IUCN Publications, New Series. No. 1. 367 pp.

1964. *The Ecology of Man in the Tropical Environment.* IUCN Publications, New Series, No. 4. 355 pp.

1969. *Eutrofication.* Proceedings of a Symposium. Washington.

1969. *Resources and Man.* National Academy of Sciences, National Research Council, San Francisco.

1970. *Environmental Quality.* The first annual report of the Council of Environmental Quality. Washington.

1971. *Environmental Quality.* The second annual report of the Council of Environmental Quality. Washington.

1971. The President's 1971 Environmental Program. Washington.

INDEX